D0556830

NEW RECREATIONS WITH MAGIC SQUARES

William H. Benson
and Oswald Jacoby

DOVER PUBLICATIONS, INC., NEW YORK

To Ann
in inadequate appreciation of
her constant encouragement and
her gracious acceptance over the years
of the role of "magic-square widow"

Published in Canada by General Publishing Company, Ltd., 30 Lesmill Road, Don Mills, Toronto, Ontario.
Published in the United Kingdom by Constable and Company, Ltd., 10 Orange Street, London WC 2.

New Recreations with Magic Squares is a new work, first published by Dover Publications, Inc., in 1976.

International Standard Book Number: 0-486-23236-0
Library of Congress Catalog Card Number: 74-28909

Manufactured in the United States of America
Dover Publications, Inc.
180 Varick Street
New York, N.Y. 10014

CONTENTS

PART III   ENUMERATION OF MAGIC SQUARES

PART IV   MATHEMATICAL PROOFS
OF THE VARIOUS METHODS

APPENDIX: Complete Listing of All Possible
Fourth-Order Magic Squares

PREFACE

"Previous experience" and a knowledge of "advanced mathematics" are by no means a prerequisite to having fun with magic squares. Quite the contrary! In fact, it is our hope that many of the readers of this book will actually have had no previous knowledge of the subject and that we can therefore be instrumental in opening for them this fascinating field of recreational mathematics.

While some of the material in Part I will be familiar to the expert, we are sure that even the most ardent magic-square enthusiast will find much of interest in the material in Parts II, III and IV, most of which are entirely original.

Part I deals with the construction of the various types of magic squares and the various added magical properties that can be included in odd and doubly-even-order squares. In Part II we develop original and very powerful methods of constructing squares that, in addition to being magic, possess other properties that can be predicted in advance. The construction of multimagic squares, including the first 32nd-order trimagic square ever to be constructed,* is also covered in this part. Part III deals with enumeration problems in general and in third-, fourth- and fifth- (pandiagonal only) order magic squares in detail. Part IV gives mathematical proofs of the methods previously employed along with further extensions thereof.

We wish to acknowledge with sincere appreciation the interest shown by, and the helpful suggestions of, the following individuals:

Mr. Francis L. Miska, Aurora, Illinois.
Mr. Herbert Coates, Milespit Hill, London, N. W. 7.
Mr. Norman Stewart, Jr., Harrisburg, Pennsylvania.
Mr. Michael O'Heeron, Carlisle, Pennsylvania.

---

*First constructed by William H. Benson in 1949.

PART I

THE WORLD
OF MAGIC SQUARES

1

INTRODUCTORY

A *magic square* may be defined as an array of numbers arranged in the form of a square so that the sum of the numbers in each row, each column and the two main diagonals total the same. This sum is known as the *magic constant*.

It follows that it takes nine numbers to form a 3-by-3 (third-order) magic square; sixteen numbers for a 4-by-4 (fourth-order) magic square; and so on up to n^2 numbers for an n-by-n (nth-order) magic square. Magic squares formed of the first n^2 natural numbers are known as *normal magic squares*. Figure 1 shows the oldest, and simplest, magic square known.

8	1	6
3	5	7
4	9	2

Fig. 1. A third-order magic square.

It is, of course, not necessary to limit the numbers to the first n^2 natural numbers. Any desired set of numbers may be used provided they have the necessary properties (as discussed later). For example, Figure 2 shows the smallest square (the magic square with the smallest magic constant) that can be constructed consisting entirely of prime numbers (counting 1 as a prime number for this purpose).

67	1	43
13	37	61
31	73	7

Fig. 2. A third-order magic square formed of prime
numbers.

The 4-by-4 magic square in Figure 3 is of interest in that it appears in the print "Melancholy," engraved by Albrecht Dürer in 1514. Notice that the two middle numbers in the bottom row give the date in which the picture was engraved.

16	3	2	13
5	10	11	8
9	6	7	12
4	15	14	1

Fig. 3. A fourth-order magic square depicted by Dürer.

Examination will show that the magic constant of 34 appears in many more places than just the rows, columns and main diagonals. The sum of the four corners is 34. In fact any number on the square and its opposite add to 17; so, the sum of any symmetrical group of four numbers on the square will be 34.

Thus there are more magic properties to this square than to the 3-by-3 square. In general, we will find that for any odd, or doubly-even (to be defined shortly) square, the larger the square the greater the possible number of magic properties.

It shouldn't take you very long to construct a normal third-order magic square (one with the numbers 1 to 9, inclusive), even if required to place the number 1 in some other spot than the middle of the top row. You will find it impossible, however, to construct such a square with the number 1 in one of the corners or with the number 5 anywhere but in the center. Placing the number 1 in the center of the first column and the number 5 in the center cell, you can construct the two squares shown in Figures 4 and 5.

6	7	2
1	5	9
8	3	4

8	3	4
1	5	9
6	7	2

Fig. 4.   Fig. 5.

It should be no harder to make a square with the nine prime numbers used in Figure 2. Again you will have to put the number 1 in the middle of a row or column, say the first column, and the number 37 in the center cell. Under these conditions you will arrive at either Figure 6 or Figure 7.

43	61	7
1	37	73
67	13	31

67	13	31
1	37	73
43	61	7

Fig. 6.   Fig. 7.

In fact it is very easy to construct normal magic squares of any odd order. You can construct them almost as fast as you can write.

It is just about as easy to construct one when *n* is 4, 8, 12, 16 etc. Such squares are known as *doubly-even magic squares*. The construction of *singly-even magic squares* (where *n* is 6, 10, 14, 18 etc.) is more difficult.

Many rules have been derived over the years for the construction of these squares, as well as for more complicated ones. In the chapters immediately following we will discuss some of the simpler methods (which are, of course, not original with the authors).

2

DOUBLY-EVEN-ORDER
MAGIC SQUARES

The construction of doubly-even-order magic squares (squares where n is divisible by 4) is exceedingly simple. The method will be demonstrated by constructing a fourth-order and an eighth-order magic square. Write the numbers 1 to 16 (1 to 64) in an auxiliary square in their natural order (see Figures 8 and 9). Now draw the dashed diagonal lines as indicated.

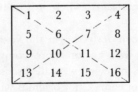

Fig. 8.

1	2	3	4	5	6	7	8
9	10	11	12	13	14	15	16
17	18	19	20	21	22	23	24
25	26	27	28	29	30	31	32
33	34	35	36	37	38	39	40
41	42	43	44	45	46	47	48
49	50	51	52	53	54	55	56
57	58	59	60	61	62	63	64

Fig. 9.

Now generate a second square similar to the original square with the exception that each number cut by one of the diagonal lines is replaced by its complement (the *complement* of a number, in this case, is that number which you must add to it to make the sum of the two numbers equal to n^2+1, n being the order of the square; see Figures 10 and 11). You will note that in Figure 10 the number 1 is replaced by its complement 16 and vice versa. Similarly, in Figure 11 the number 1 is replaced by its complement 64 and vice versa, and so on.

16	2	3	13
5	11	10	8
9	7	6	12
4	14	15	1

Fig. 10. A symmetrical fourth-order magic square.

The probability is that the Dürer square in Figure 3 was constructed in this fashion, and that then the second and third columns were interchanged (which does not affect the magical properties) in order to make the numbers in the middle of the bottom row agree with the year 1514.

This method of construction can be applied to magic squares of any doubly-even order. All that is necessary is to divide the natural-

64	2	3	61	60	6	7	57
9	55	54	12	13	51	50	16
17	47	46	20	21	43	42	24
40	26	27	37	36	30	31	33
32	34	35	29	28	38	39	25
41	23	22	44	45	19	18	48
49	15	14	52	53	11	10	56
8	58	59	5	4	62	63	1

Fig. 11. A symmetrical eighth-order magic square.

order generating square into 4-by-4 subsquares, draw the dotted diagonal lines, and then replace each number cut by the diagonals of these subsquares by its complement. You have probably already noticed that magic squares constructed by this method have the property that any two numbers symmetrically located with respect to the center of the square are complementary. Magic squares which possess this property are said to be *symmetrical magic squares*.

By their very nature, symmetrical magic squares lend themselves readily to rearrangement. If you transpose any two rows or columns of an ordinary magic square the chances are that the main diagonals of the resultant square will be incorrect. However, in the case of symmetrical magic squares the property of symmetry makes it possible to interchange any two rows or columns which are equally distant from the center of the square and still have a symmetrical magic square. Figure 3 is an example of this. It is identical with Figure 10, with columns 2 and 3 reversed.

If you are careful, you can take considerable liberty with symmetrical doubly-even-order magic squares with rather astonishing results. For example, if you take Figure 11 and reverse the last half of the columns (that is, columns 5 and 8 and columns 6 and 7) you get Figure 12a, which, while magic, is not symmetrical.

Suppose you now reverse the last half of the rows of Figure 12a. This will result in Figure 12b, which has the remarkable property that not only are the rows, columns and main diagonals correct, but the broken diagonals are also correct. For example:

$$8 + 26 + 46 + 12 + 57 + 39 + 19 + 53 = 260, \text{ also}$$
$$17 + 26 + 59 + 52 + 48 + 39 + 6 + 13 = 260, \text{ etc.}$$

64	2	3	61	57	7	6	60
9	55	54	12	16	50	51	13
17	47	46	20	24	42	43	21
40	26	27	37	33	31	30	36
32	34	35	29	25	39	38	28
41	23	22	44	48	18	19	45
49	15	14	52	56	10	11	53
8	58	59	5	1	63	62	4

Fig. 12a.

64	2	3	61	57	7	6	60
9	55	54	12	16	50	51	13
17	47	46	20	24	42	43	21
40	26	27	37	33	31	30	36
8	58	59	5	1	63	62	4
49	15	14	52	56	10	11	53
41	23	22	44	48	18	19	45
32	34	35	29	25	39	38	28

Fig. 12b. A pandiagonal eighth-order magic square.

Magic squares which have this property are known as *pandiagonal* (sometimes as *panmagic*, *continuous*, *perfect*, *diabolic*, *Nasik* or *Jaina*) *magic squares*.

As another example of what can be done by transposing columns and rows, let us rearrange the columns of Figure 11 as shown in Figure 13. This square, in which the columns are transposed in the order 1-6-8-3-4-7-5-2, is not even magic, but notice what happens when we transpose the rows in a similar manner.

We now have Figure 14, a truly remarkable magic square—formed by the simple process of transposing the columns and rows of Figure 11 in the above-listed manner—which possesses the following properties:

(1) It meets *all* the requirements of Franklin's square (see note at end of chapter).

(2) The main 8-by-8 square is pandiagonal, that is, the sum of the eight numbers in each row, column, main diagonal and broken diagonal equals 260.

64	6	57	3	61	7	60	2
9	51	16	54	12	50	13	55
17	43	24	46	20	42	21	47
40	30	33	27	37	31	36	26
32	38	25	35	29	39	28	34
41	19	48	22	44	18	45	23
49	11	56	14	52	10	53	15
8	62	1	59	5	63	4	58

Fig. 13.

64	6	57	3	61	7	60	2
41	19	48	22	44	18	45	23
8	62	1	59	5	63	4	58
17	43	24	46	20	42	21	47
40	30	33	27	37	31	36	26
49	11	56	14	52	10	53	15
32	38	25	35	29	39	28	34
9	51	16	54	12	50	13	55

Fig. 14. A pandiagonal eighth-order magic square with many additional magical properties.

(3) Each of the 4-by-4 subsquares enclosed in heavy lines is also pandiagonal!

It is truly a remarkable magic square!

While the above method of constructing doubly-even-order magic squares is the simplest we know, nevertheless there are only a limited number of such squares that can be constructed in this manner. We will, therefore, need to return to this subject later on in the book.

FRANKLIN'S SQUARE

The ancients were interested in magic squares, engraving them on metal or stone, and apparently used them as amulets or talismans. Even the great Benjamin Franklin played with these squares. Figure 15 shows one that he constructed.

52	61	4	13	20	29	36	45
14	3	62	51	46	35	30	19
53	60	5	12	21	28	37	44
11	6	59	54	43	38	27	22
55	58	7	10	23	26	39	42
9	8	57	56	41	40	25	24
50	63	2	15	18	31	34	47
16	1	64	49	48	33	32	17

Fig. 15.   Benjamin Franklin's "square of 8."

This square, while not magic in the true sense (the main diagonals are incorrect) has the following properties:

(1) Every straight row (or column) of eight numbers added together makes 260 and half of each row (or column) makes 130.

(2) The bent row of eight numbers, ascending and descending diagonally (that is, ascending from 16 to 10 and descending from 23 to 17), and each of its parallel bent rows of eight numbers, added together make 260. Also, the bent row descending from 52 to 54 and ascending from 43 to 45, and each of its parallel bent rows of eight numbers, added together make 260. Similarly bent rows formed from either side total 260.

(3) The sum of any four number which form a 2-by-2 subsquare equals 130. One example is shown in heavy lines: $28 + 37 + 38 + 27 = 130$.

(4) In fact, the sum of any four numbers symmetrically located with respect to the intersection of any line separating two columns and any line separating two rows equals 130. For example, the sum of the four circled numbers $61 + 20 + 6 + 43 = 130$.

3

ODD-ORDER MAGIC SQUARES

One of the oldest and simplest of the many ways for constructing magic squares of any odd order (squares where n is odd) is known as the De la Loubère's rule. Specifically it consists of starting with the number 1 in the middle cell of the top row and proceeding diagonally upward to the right until you are blocked. At this point you drop down one cell and start the next series of numbers from this point, proceeding diagonally upward to the right until blocked, when you again drop down one cell. You continue this process until the square is completed.

Let us apply this rule, which is not nearly as complicated as it sounds, to the construction of a fifth-order magic square. In completing any given diagonal, consider the square as continuous, that is, that the bottom row is also the next row above the top row and the left-hand column is also the next column to the right of the right-hand column. Until you have become familiar with this cyclic concept it will probably be an aid to draw an auxiliary row and column as shown in Figure 16a.

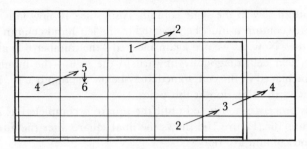

Fig. 16a.

In accordance with the rule, the number 1 is placed in the middle cell of the top row. The number 2 is now placed in the cell one space diagonally up to the right (that is, one row up and one column to the right). Since this cell is in the auxiliary row (and not in the square proper) it is repeated in the bottom row in the same column. The number 3 is written in its normal position in the next cell diagonally upward to the right. Likewise the number 4 is written in the next diagonal cell and, since this number falls in an auxiliary column, it is repeated in the same row in the left-hand column. The number 5 is written in the cell diagonally upward to the right from 4. At this point you are blocked for the first time. The number 6 cannot be placed in the cell diagonally upward to the right—the number 1 is already in that cell—and must, therefore, be placed in the cell immediately below the number 5. Figure 16a shows the appearance of the square at this point.

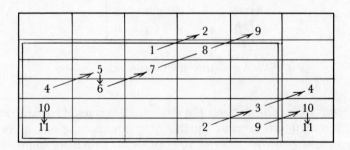

Fig. 16b.

The second diagonal is now started and the numbers 7, 8 and 9 written in place. Since 9 falls in the auxiliary row it is repeated in the bottom row in the same column. Also, since 10 now falls in the auxiliary column it must be repeated in the left-hand column in the same row. Now, as we are again blocked—the number 6 is already in the next cell, diagonally upward to the right—the number 11 must be placed in the cell immediately below the number 10. Figure 16b shows the situation at this point.

This process is continued until the square is completed as shown in Figure 16c. Figure 16d shows the final square after the auxiliary row and column are removed.

	18	25	2	9	11
17	24	1	8	15	17
23	5	7	14	16	23
4	6	13	20	22	4
10	12	19	21	3	10
11	18	25	2	9	11

Fig. 16c.

17	24	1	8	15
23	5	7	14	16
4	6	13	20	22
10	12	19	21	3
11	18	25	2	9

Fig. 16d.   A symmetrical fifth-order magic square.

Figures 17a, 17b, 17c and 17d show similar stages of construction for a seventh-order magic square. After constructing one or two magic squares by this method, you can hardly help being impressed by its simplicity. In fact, by its use anyone can construct a magic square of any odd order almost as fast as he can write. Note also that all magic squares constructed by this method are symmetrical.

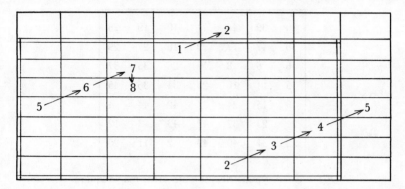

Fig. 17a.

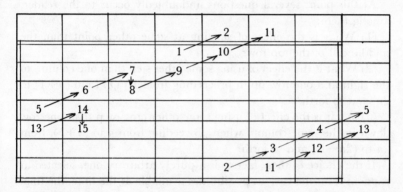

Fig. 17b.

	31	40	49	2	11	20	22
30	39	48	1	10	19	28	30
38	47	7	9	18	27	29	38
46	6	8	17	26	35	37	46
5	14	16	25	34	36	45	5
13	15	24	33	42	44	4	13
21	23	32	41	43	3	12	21
22	31	40	49	2	11	20	22

Fig. 17c.

30	39	48	1	10	19	28
38	47	7	9	18	27	29
46	6	8	17	26	35	37
5	14	16	25	34	36	45
13	15	24	33	42	44	4
21	23	32	41	43	3	12
22	31	40	49	2	11	20

Fig. 17d. A symmetrical seventh-order magic square.

At this point several questions undoubtedly occur to the reader, among them:

(1) What is the effect of starting at some other point than the middle cell of the top row?

(2) What is the effect of using some other step than one column to the right and one row up in proceeding from 1 to n, from $(n+1)$ to $2n$, and so forth?

(3) What is the effect of using some other cross-step than one row down in the same column when proceeding from n to $(n+1)$, from $2n$ to $(2n+1)$, and so forth?

If the choice of these factors is a purely arbitrary one, selected at random from the various possibilities available at the time, the final result will vary considerably. In Figure 18 the construction was begun by placing the number 1 in the cell immediately above the center cell, the method being unchanged in other particulars. It will be noted that while this square gives the correct totals for all the rows and columns, it gives the wrong total for one of the main

11	18	25	2	9
17	24	1	8	15
23	5	7	14	16
4	6	13	20	22
10	12	19	21	3

Fig. 18. A semimagic square.

diagonals. Such squares (those with correct rows and columns, but with one or both diagonals incorrect) are known as *semimagic squares*.

23	6	19	2	15
10	18	1	14	22
17	5	13	21	9
4	12	25	8	16
11	24	7	20	3

Fig. 19. A symmetrical magic square.

In Figure 19 the construction was begun by placing the number 1 in the cell immediately above the center cell (as in the case of Figure 18) and the method was unchanged in other particulars except that when blocked at the end of any given diagonal the next number was placed in the same column two rows up rather than one row down. The resultant square is a symmetrical magic square. This is known as the Barchet de Méziriac method.

In Figure 20 the construction was made in a normal manner except that when blocked the next move was one cell to the right. This square is not even semimagic. The totals of the columns are incorrect.

Figure 21 was constructed entirely in the normal manner except that instead of moving one column to the right and one row up along the upward diagonal, a "knight's move" of one column to the right and two rows up was followed. The resultant square is not only symmetrical, it is also pandiagonal!

19	25	1	7	13
24	5	6	12	18
4	10	11	17	23
9	15	16	22	3
14	20	21	2	8

Fig. 20. Nonmagic.

10	18	1	14	22
11	24	7	20	3
17	5	13	21	9
23	6	19	2	15
4	12	25	8	16

Fig. 21.   A symmetrical, pandiagonal magic square.

As in the case of doubly-even magic squares we shall return to this discussion later when we take up methods of construction of magic squares with additional properties (which can be predicted in advance).

4

SINGLY-EVEN-ORDER
MAGIC SQUARES

There is no simple method of constructing singly-even-order magic squares (squares where n is divisible by 2, but not by 4). The following method is an original method based upon that of Ralph Strachey.*

In a singly-even-order square, n is of the form $2(2m+1)$ where m is any positive integer. Let $k = n^2/4$ and construct the following table for future reference:

m	n	n^2	k	$2k$	$3k$
1	6	36	9	18	27
2	10	100	25	50	75
3	14	196	49	98	147
4	18	324	81	162	243
5	22	484	121	242	363
etc.					

---

*A generalization by the authors of Ralph Strachey's method as given in *Mathematical Recreations and Essays* by W. W. R. Ball, The Macmillan Company, pp. 196–99.

We will demonstrate the method by constructing a tenth-order magic square. In this case we have: $m = 2$, $n = 10$, $k = 25$, $2k = 50$ and $3k = 75$.

The first step consists in thinking of the square as being divided into four equal quarters as shown in Figure 22a.

A	B
C	D

Fig. 22a.

In any quarter, say A, write the number $3k$ (75 in the case of our tenth-order example) in m (equal to 2 in this case) cells in each row. The cells may be selected at random, but you must place $m + 1$ (here equal to 3) of them in cells on the main diagonal (of the entire tenth-order square). Write 0 in the remaining cells of this quarter. The situation now appears as shown in Figure 22b.

75	0	0	0	75					
0	75	0	0	75					
75	0	0	75	0			B		
75	0	75	0	0					
0	0	0	75	75					
		C					D		

Fig. 22b.

Now select the adjacent quarter (above or below as the case may be), in this case C, and enter 0 and 75 in the cells in the reverse manner. That is, if the number in the first cell in the top row is 75, the number in the same column in the bottom row of C will be 0; if the number in the xth cell in the yth row from the top in quarter A is 75 (or 0), the number in the same column in the yth row from the bottom in quarter C will be 0 (or 75). The situation now appears as shown in Figure 22c.

75	0	0	0	75					
0	75	0	0	75					
75	0	0	75	0		B			·
75	0	75	0	0					
0	0	0	75	75					
75	75	75	0	0					
0	75	0	75	75					
0	75	75	0	75		D			
75	0	75	75	0					
0	75	75	75	0					

Fig. 22c.

The next quarter to be filled in is the remaining one adjacent to the quarter in which we started, B in this case. In this quarter write $2k$ (equal to 50 in our example) in $m + 2$ (here equal to 4) cells in each row. Here again you may place this number, 50, in any of the cells you desire, except that you must make sure that exactly $m + 2$ (equal to 4) of them are in cells that lie on the main diagonal. Write k (25 in this case) in the remaining cells in this quarter. This leaves only one quarter to be completed. Just as the second quarter (C in this example) was filled with the reverse of the first quarter, so the fourth quarter is filled with the reverse of the third. That is, if the number in the xth column of the yth row from the top in quarter B is 50 (or 25), the number in the same column in the yth row from the bottom in quarter D will be 25 (or 50). The completed preliminary construction square is shown in Figure 22d.

75	0	0	0	75	50	50	25	50	50
0	75	0	0	75	50	50	50	50	25
75	0	0	75	0	25	50	50	50	50
75	0	75	0	0	50	25	50	50	50
0	0	0	75	75	50	25	50	50	50
75	75	75	0	0	25	50	25	25	25
0	75	0	75	75	25	50	25	25	25
0	75	75	0	75	50	25	25	25	25
75	0	75	75	0	25	25	25	25	50
0	75	75	75	0	25	25	50	25	25

Fig. 22d.

We have now constructed a (nonnormal) magic square of the tenth order using only the numbers 0, 25, 50 and 75. The sum of each row, column and main diagonal is 375. Now construct a normal magic square of the $n/2$ (5 in this case) order using any method that you wish. For the purposes of this example we will use the magic square shown in Figure 21. Superimpose this square upon the first and third quarters (A and B as we chose them); by superimposition is meant adding the numbers in the corresponding positions to each other (see Figure 22e). In a similar manner superimpose the mirror of this magic square in the remaining two quarters. Notice that in the final result, as shown in Figure 22e, any given number in the 5-by-5 magic square is paired once—and only once—with 0, 25, 50 and 75. This means that there will be no duplication and that the final square, Figure 22f, will be formed of the numbers 1 to 100 inclusive.

Now construct Figure 22f by adding the component numbers in each cell of Figure 22e. Since each set of numbers forms a (nonnormal) tenth-order magic square, their sum will also form a magic square and, since it is formed of the numbers 1 to 100 inclusive, it will also be normal.

You may have wondered why no example of a symmetrical, or of a pandiagonal, singly-even magic square is given. The answer is quite simple. As will be shown later, they do not exist!

75+10	0+18	0+1	0+14	75+22	50+10	50+18	25+1	50+14	50+22
0+11	75+24	0+7	0+20	75+3	50+11	50+24	50+7	50+20	25+3
75+17	0+5	0+13	75+21	0+9	25+17	50+5	50+13	50+21	50+9
75+23	0+6	75+19	0+2	0+15	50+23	25+6	50+19	50+2	50+15
0+4	0+12	0+25	75+8	75+16	50+4	25+12	50+25	50+8	50+16
75+4	75+12	75+25	0+8	0+16	25+4	50+12	25+25	25+8	25+16
0+23	75+6	0+19	75+2	75+15	25+23	50+6	25+19	25+2	25+15
0+17	75+5	75+13	0+21	75+9	50+17	25+5	25+13	25+21	25+9
75+11	0+24	75+7	75+20	0+3	25+11	25+24	25+7	25+20	50+3
0+10	75+18	75+1	75+14	0+22	25+10	25+18	50+1	25+14	25+22

Fig. 22e.

85	18	1	14	97	60	68	26	64	72
11	99	7	20	78	61	74	57	70	28
92	5	13	96	9	42	55	63	71	59
98	6	94	2	15	73	31	69	52	65
4	12	25	83	91	54	37	75	58	66
79	87	100	8	16	29	62	50	33	41
23	81	19	77	90	48	56	44	27	40
17	80	88	21	84	67	30	38	46	34
86	24	82	95	3	36	49	32	45	53
10	93	76	89	22	35	43	51	39	47

Fig. 22f.   A tenth-order magic square.

Analysis will show that this method, which (as modified by the authors) consists essentially of superimposing a normal magic square (of order $n/2$) upon a nonnormal magic square (of order n) which is formed of the numbers 0, $n^2/4$, $n^2/2$ and $3n^2/4$ in such a manner that it has the property of being symmetrical about the horizontal line separating the upper half from the lower half, is perfectly general. Not only can you use it to construct magic squares of any order n where $n = 2(2m + 1)$, but—with certain very obvious modifications—it can be extended to cover the construction of magic squares of any order n where $n = 4m$.

For example, Figure 23a is an eighth-order (nonnormal) pandiagonal magic square formed with 0 and 48 (each used an equal number of times in each row and in the main diagonal) in quadrant A and 16 and 32 (each used an equal number of times in each row and in the main diagonal) in quadrant B. Also, as is necessary, the lower half is symmetrical to the upper half about the horizontal line separating the two halves.

0	48	48	0	32	16	16	32
48	0	0	48	16	32	32	16
0	48	48	0	32	16	16	32
48	0	0	48	16	32	32	16
0	48	48	0	32	16	16	32
48	0	0	48	16	32	32	16
0	48	48	0	32	16	16	32
48	0	0	48	16	32	32	16

Fig. 23a.

For the fourth-order magic square that we need to complete our eighth-order magic square let us take Figure 10 and transpose the third and fourth rows and the third and fourth columns. This will give us the pandiagonal magic square shown in Figure 23b.

16	2	13	3
5	11	8	10
4	14	1	15
9	7	12	6

Fig. 23b.

Figure 23c is Figure 23a with Figure 23b superimposed in quadrants *A* and *B* and mirrored in quadrants *C* and *D*. Again notice that this results in any given number in Figure 23b being combined in Figure 23c with each of the numbers 0, 16, 32 and 48 once, and only once, thus ensuring that the final square will be composed of the numbers 1 to 64, inclusive.

0+16	48+2	48+13	0+3	32+16	16+2	16+13	32+3
48+5	0+11	0+8	48+10	16+5	32+11	32+8	16+10
0+4	48+14	48+1	0+15	32+4	16+14	16+1	32+15
48+9	0+7	0+12	48+6	16+9	32+7	32+12	16+6
0+9	48+7	48+12	0+6	32+9	16+7	16+12	32+6
48+4	0+14	0+1	48+15	16+4	32+14	32+1	16+15
0+5	48+11	48+8	0+10	32+5	16+11	16+8	32+10
48+16	0+2	0+13	48+3	16+16	32+2	32+13	16+3

Fig. 23c.

All that is left to do to get our final square, the pandiagonal magic square shown in Figure 23d, is to add together the two numbers in each cell.

16	50	61	3	48	18	29	35
53	11	8	58	21	43	40	26
4	62	49	15	36	30	17	47
57	7	12	54	25	39	44	22
9	55	60	6	41	23	28	38
52	14	1	63	20	46	33	31
5	59	56	10	37	27	24	42
64	2	13	51	32	34	45	19

Fig. 23d.   An eighth-order pandiagonal magic square.

5

CONCENTRIC (OR BORDERED) MAGIC SQUARES

A *concentric (or bordered) magic square* is one that will remain magic if the borders (consisting of the top and bottom rows and the right-hand and left-hand columns) are removed—one border at a time. Squares of any odd order can be constructed by starting with a third-order magic square, and squares of any even order by starting with a fourth-order magic square. Figure 24 is an example of this type of square.

19	2	20	1	23
4	16	9	14	22
18	11	13	15	8
21	12	17	10	5
3	24	6	25	7

Fig. 24. A fifth-order bordered magic square.

While the method employed in constructing this square is more or less self-evident, let us examine it more closely. The original nucleus is clearly the third-order magic square shown in Figure 1 (repeated here for convenience).

8	1	6
3	5	7
4	9	2

If we wish to add a border to this magic square, and thus convert it into a fifth-order bordered magic square, we first add $2(n-1)$ $=2(5-1)=8$ to each number in the nucleus square. This will now become a (nonnormal) magic square formed of the numbers 9 to 17, inclusive, with a magic constant of $(n-2)(n^2+1)/2$. We now have available for the construction of the border the eight pairs of numbers 1-25, 2-24, 3-23, 4-22, 5-21, 6-20, 7-19 and 8-18. Notice that each of these pairs totals $(n^2+1)=26$ and that if we add any given pair to any row, column or main diagonal of the nucleus square, the total for that row, column or main diagonal will have the correct value since $(n^2+1)+(n-2)(n^2+1)/2=n(n^2+1)/2$. All that remains to be done is to assign these pairs so that the totals of the outer (or border) rows and columns are correct. That this is possible is demonstrated by Figure 24.

In constructing such magic squares, however, it is not necessary to resort to trial and error. Any given odd-order magic square can have a magic border added to it by following the process described below. While this process is complicated to describe, actually it is—as the example will show—very easy to use. Draw a blank n-by-n square of the desired order, add $2(n-1)$ to each number in the initial $(n-2)$-by-$(n-2)$ magic square, and insert it in the center of the blank n-by-n square. If we desire to construct a fifth-order bordered magic square using Figure 1 as a nucleus, the result so far would look like Figure 25a.

Next, place in the corners of your new square the numbers indicated in Figure 26 on page 29, which shows the mathematical basis of this type of magic square. That is, place $n^2-(n-1)/2$ $=25-(5-1)/2=23$ in the upper left-hand corner; $n^2-3(n-1)/2$ $=25-3(5-1)/2=19$ in the upper right-hand corner; $n+(n-1)/2$

	16 9 14	
	11 13 15	
	12 17 10	

23		19
	16 9 14	
	11 13 15	
	12 17 10	
7		3

Fig. 25a.   *Fig. 25b.*

$= 5 + (5-1)/2 = 7$ in the lower left-hand corner; and $1 + (n-1)/2$ $= 1 + (5-1)/2 = 3$ in the lower right-hand corner. Your new square now looks like Figure 25b.

Next, fill in the remaining blank cells in the left-hand column with the numbers listed in Set A (below) and the remaining blank cells in the bottom row with the numbers listed in Set B. In generating the actual numbers in each set the letters $a, b, c, \ldots, r$, $a', b', c', \ldots, r'$ are arbitrarily assigned one of the values $1, 2, 3, \ldots, (n-2)$, each of these values being used once, and only once, with the exception of the number $(n-1)/2$ (which you will notice has already been used in generating the corner numbers).

SET A

The number n and also $(n-3)/2$ terms from each of the following columns:

$$n^2 - a \qquad n + a$$
$$n^2 - b \qquad n + b$$
$$\vdots \qquad \vdots$$
$$n^2 - r \qquad n + r$$

SET B

The number n^2 and also $(n-3)/2$ terms from each of the following columns:

$$n^2 - a' \qquad n + a'$$
$$n^2 - b' \qquad n + b'$$
$$\vdots \qquad \vdots$$
$$n^2 - r' \qquad n + r'$$

Remembering that $n = 5$ in our example, we see that we have the numbers 1 and 3 available for our use in generating Sets A and B. Say we use 1 in Set A and 3 in Set B. Then: Set A becomes $\rightarrow n = 5$; $n^2 - a = 25 - 1 = 24$; and $n + a = 5 + 1 = 6$; and Set B becomes $\rightarrow n^2 = 25$; $n^2 - a' = 25 - 3 = 22$; and $n + a' = 5 + 3 = 8$.

The order in which the actual numbers are assigned to the blank cells is immaterial, the only restriction being that the numbers from Set A are used to fill the left-hand column and the numbers from Set B are used to fill the bottom row. Having done this, all that remains is to write the complements of the numbers in the left-hand column in the right-hand column (being careful to write any given number and its complement in the same row, and to complete the top row in a similar fashion with the complements of the numbers in the bottom row. Figure 25c is the final result.

23	1	4	18	19
5	16	9	14	21
24	11	13	15	2
6	12	17	10	20
7	25	22	8	3

Fig. 25c. A fifth-order bordered magic square.

$n^2-(n-1)/2$	Complements of Set B	$n^2-3(n-1)/2$
Set A	Original odd-order magic square plus $2(n-1)$	Complements of Set A
$n+(n-1)/2$	Set B	$1+(n-1)/2$

Fig. 26.

The construction of even-order bordered magic squares is a similar process, except that the rules you must follow depend upon whether the magic square that you wish to border is a singly-even-order or a doubly-even-order square. We shall first construct a sixth-order bordered magic square by using Figure 3 as a nucleus and the basic rules as given in Figure 27.

n^2	Complements of Set D	n^2+1-n
Set C	Doubly-even magic square plus $2(n-1)$	Complements of Set C
n	Set D	1

Fig. 27.

SET C

This set consists of (n^2-n-1), (n^2-n-3), $(n+3)$ and (4) along with $(n-6)/4$ terms from each of the columns in Set E.

SET D

This set consists of (n^2-1), (n^2-2), (n^2-n) and (5) along with $(n-6)/4$ terms from each of the columns listed in Set F.

SET E

This set consists of $n/4$ terms from each of the following columns (note: the left-hand corner numbers are included in these $n/4$ terms):

$$
\begin{array}{cccc}
n^2-a & n+a & n^2+1-n-p & 1+p \\
n^2-b & n+b & n^2+1-n-q & 1+q \\
\vdots & \vdots & \vdots & \vdots \\
n^2-k & n+k & n^2+1-n-w & 1+w
\end{array}
$$

SET F

This set consists of (n^2-x) and $(n^2+1-n-x)$ along with $(n-4)/4$ terms from each of the following columns:

$$
\begin{array}{cccc}
n^2-a' & n+a' & n^2+1-n-p' & 1+p' \\
n^2-b' & n+b' & n^2+1-n-q' & 1+q' \\
\vdots & \vdots & \vdots & \vdots \\
n^2-k' & n+k' & n^2+1-n-w' & 1+w'
\end{array}
$$

Here $a, b, \ldots, k, p, q, \ldots, w, a', b', \ldots, k', p', q', \ldots, w',$ and x are assigned one of the values $0, 1, 2, \ldots, (n-2)$, each of these values being assigned once, and only once, with the exception that in the case of sets C and D the values 0, 1, 2, 3 and 4 shall not be assigned as they have already been used in forming the corner terms and those listed as specific members of these sets. In addition, in the case of Sets E and F it is necessary for the values of a, b and x to be so chosen that $2x = a + b$.

Figure 28 shows the sixth-order bordered magic square constructed by following the above instructions.

36	2	3	7	32	31
29	26	13	12	23	8
27	15	20	21	18	10
9	19	16	17	22	28
4	14	25	24	11	33
6	35	34	30	5	1

Fig. 28.

Figure 29 gives the basic instructions for forming a border around a singly-even magic square.

$n^2 - b$	Complements of Set F	$n^2 + 1 - n - a$
Set E	Singly-even magic square plus $2(n-1)$	Complements of Set E
$n + a$	Set F	$1 + b$

Fig. 29.

Using a sixth-order bordered magic square as a nucleus, and the rules contained in Figure 29, you can construct an eighth-order bordered magic square. Figure 30 is such a square. In its construction Figure 28 was used as a nucleus, along with $a = 1$, $b = 3$, $x = (1 + 3)/2 = 2$, $p = 0$, $q = 4$, $a' = 5$ and $p' = 6$.

61	3	10	6	52	14	58	56
63	50	16	17	21	46	45	2
11	43	40	27	26	37	22	54
57	41	29	34	35	32	24	8
53	23	33	30	31	36	42	12
1	18	28	39	38	25	47	64
5	20	49	48	44	19	15	60
9	62	55	59	13	51	7	4

Fig. 30.

It should not be assumed that the above method is the only way to create bordered magic squares. In fact, a 3-by-3 center square for odd-order bordered magic squares can easily be formed by assigning arbitrary values to a and b in Figure 31 subject to the following restrictions:

$$a < b < (n^2 + 1)/2; \qquad (n^2 + 1)/2 < (a + b);$$

$$(n^2 + 1)/2 \neq (2b - a).$$

(Here the symbol $<$ indicates "is less than" and $\neq$ indicates "is not equal to.")

b	$[a - b] + [(n^2 + 1)/2]$	$[(n^2 + 1) - a]$
$[3(n^2 + 1)/2] - [a + b]$	$[(n^2 + 1)/2]$	$[a + b] - [(n^2 + 1)/2]$
a	$[b - a] + [(n^2 + 1)/2]$	$[(n^2 + 1) - b]$

Fig. 31.

Figure 32 is a seventh-order bordered magic square built about Figure 31 with $a = 3$ and $b = 24$. Note that the number in the center cell is $[(n^2+1)/2]$ and that the square is constructed by using pairs of numbers that add up to (n^2+1), that is, their value is based upon the order of the final square.

22	41	34	27	17	5	29
1	35	6	42	11	31	49
38	10	24	4	47	40	12
37	18	48	25	2	32	13
36	43	3	46	26	7	14
20	19	44	8	39	15	30
21	9	16	23	33	45	28

Fig. 32.

As a matter of possible interest, it is fairly easy to show that the number of such 3-by-3 nucleus squares that can be formed is:

$(3n^4 - 26n^2 + 23)/48$ when n is prime to 3, and

$(3n^4 - 26n^2 + 39)/48$ when n is not prime to 3.

When you also take into account the numerous variations within each border, the number of bordered magic squares for even a reasonable size n staggers the imagination.

Before leaving the subject of bordered magic squares, it should be noted that the methods described in this chapter furnish still another way to construct magic squares of various orders.

6

COMPOSITE MAGIC
SQUARES

Very little study is needed to see how simple it is to construct a magic square of mn (m times n) order when you know magic squares of the m-order and of the n-order.

Divide your mn square into n^2 subsquares, each being m-by-m. In each cell of these subsquares put m^2 times one less than the corresponding number in the nth-order square. Let us start with the magic squares shown in Figures 1 and 3. This will give us Figure 33, a (nonnormal) twelfth-order magic square.

112	112	112	112	0	0	0	0	80	80	80	80
112	112	112	112	0	0	0	0	80	80	80	80
112	112	112	112	0	0	0	0	80	80	80	80
112	112	112	112	0	0	0	0	80	80	80	80
32	32	32	32	64	64	64	64	96	96	96	96
32	32	32	32	64	64	64	64	96	96	96	96
32	32	32	32	64	64	64	64	96	96	96	96
32	32	32	32	64	64	64	64	96	96	96	96
48	48	48	48	128	128	128	128	16	16	16	16
48	48	48	48	128	128	128	128	16	16	16	16
48	48	48	48	128	128	128	128	16	16	16	16
48	48	48	48	128	128	128	128	16	16	16	16

Fig. 33.

Clearly, if we add Figure 3 to each of these subsquares we will
have a twelfth-order magic square formed of the numbers 1 to 144,
inclusive. Figure 34 is the final result.

128	115	114	125	16	3	2	13	96	83	82	93
117	122	123	120	5	10	11	8	85	90	91	88
121	118	119	124	9	6	7	12	89	86	87	92
116	127	126	113	4	15	14	1	84	95	94	81
48	35	34	45	80	67	66	77	112	99	98	109
37	42	43	40	69	74	75	72	101	106	107	104
41	38	39	44	73	70	71	76	105	102	103	108
36	47	46	33	68	79	78	65	100	111	110	97
64	51	50	61	144	131	130	141	32	19	18	29
53	58	59	56	133	138	139	136	21	26	27	24
57	54	55	60	137	134	135	140	25	22	23	28
52	63	62	49	132	143	142	129	20	31	30	17

Fig. 34.   A twelfth-order magic square.

7

A PRIME CURIOSITY

The question "Can you make a magic square using only prime numbers?" is likely to arise in the mind of anyone interested in numbers. The answer is, of course, "Yes." Figure 2 is an example of such a square.

It remained for J. N. Muncey of Jessup, Iowa, to show (in 1913) that the smallest magic square that can be constructed of *consecutive* odd prime numbers (counting 1 as a prime number) is one of the twelfth order.* This remarkable square is reproduced here in Figure 35 as a fine example of an unusual nonnormal magic square. Incidentally, it is quite a simple matter to prove that it is not possible to construct such a square of any order smaller than twelve.

**Scientific American* Magazine, published by Scientific American, Inc., New York, Vol. 210, no. 3, pp. 126–7.

1	823	821	809	811	797	19	29	313	31	23	37
89	83	211	79	641	631	619	709	617	53	43	739
97	227	103	107	193	557	719	727	607	139	757	281
223	653	499	197	109	113	563	479	173	761	587	157
367	379	521	383	241	467	257	263	269	167	601	599
349	359	353	647	389	331	317	311	409	307	293	449
503	523	233	337	547	397	421	17	401	271	431	433
229	491	373	487	461	251	443	463	137	439	457	283
509	199	73	541	347	191	181	569	577	571	163	593
661	101	643	239	691	701	127	131	179	613	277	151
659	673	677	683	71	67	61	47	59	743	733	41
827	3	7	5	13	11	787	769	773	419	149	751

Fig. 35.

PART II

THE WIDENING HORIZONS OF MAGIC SQUARES

8

INTRODUCTORY

In Part I we were interested in the simplest possible methods for constructing magic squares of various orders and types and in becoming familiar with some of the added properties that could be included in odd and doubly-even-order magic squares. In this part we are going to be interested in flexibility and in the development of methods allowing us to construct squares that, in addition to being magic, possess additional properties which can be predicted in advance.

You will have noticed the simplicity of the De la Loubère and Méziriac methods and you may have wondered why they, or some other cyclical method, haven't been extended to even-order magic squares. Without knowing, of course, we believe the answer lies in the fact that, as Ball proves in his paragraph on "Generalization of De la Loubère's Rule,"* it is impossible to use any such method for

---

*W. W. R. Ball, *op. cit.* pp. 204–207.

the direct construction of even-order magic squares. However the authors, by the combination of an original generalization of De la Loubère's rule with an entirely new (as far as we can determine) concept, have developed a cyclical method for constructing magic squares. This method, in addition to greatly increasing the flexibility of the final square, is believed to be the first cyclical method to be developed that is applicable to both odd-order and even-order magic squares.

9

A NEW APPROACH*

As a first step let us identify the individual cells forming the square by the coordinate (x,y) where x and y equal $0,1,2,\ldots,(n-1)$ counting from the left to the right and from the bottom up. Figure 36a gives the coordinates when n equals 5.

0,4	1,4	2,4	3,4	4,4
0,3	1,3	2,3	3,3	4,3
0,2	1,2	2,2	3,2	4,2
0,1	1,1	2,1	3,1	4,1
0,0	1,0	2,0	3,0	4,0

Fig. 36a.

*This method is (as far as known) original with the authors.

Since we shall be using cyclical steps, all coordinates that differ by an exact multiple of n are equivalent. In a fifth-order square this means simply that 14, 9, 4, -1, -6, -11 etc. are all equivalent. As another example, in a seventh-order square 21, 14, 7, 0, -7, -14 etc., are all equivalent. Thus the cell $(14, -11)$ would be equivalent to the cell $(4,4)$ in a fifth-order square and to $(0,3)$ in a seventh-order square. More generally we see that if j and k are integers (positive or negative and not necessarily different) then:

$$(jn + x, kn + y) \text{ is equivalent to } (x,y).$$

The second step is to select four integers C, R, c and r (positive or negative and not necessarily different) such that:

(1) Each one is greater than minus n and less than plus n.

(2) None are equal to zero.

(3) The difference $(Cr - cR)$ is prime to n. (Two integers are prime to each other when they have no common factor other than 1; it follows that 1 is prime to all other integers.)

In order to demonstrate our method let us consider the following two combinations of values for C, R, c and r:

	Case 1	Case 2
C	$+2$	$+1$
R	$+1$	$+1$
c	-1	-1
r	-2	-2
$(Cr - cR)$	-3	-1

The construction, once the various constants have been selected, is actually quite simple. The easiest way to explain it is by example. We shall first construct magic squares of the fifth, eighth and tenth order using the values of C, R, c and r given in Case 1. Note particularly that these values, as well as $(Cr - cR)$, meet the above three requirements.

A unique feature of our method is that a very general intermediate square is first constructed by the aid of n series of pairs of letters (one of each pair being a capital letter and the other a lower-case letter), namely, the A series, the B series,..., the N series. Each series in turn contains the n lower-case letters $a, b, c,..., n$.

The first step in the actual construction is to generate this intermediate square by the use of one series of cyclical steps after another. Each individual cell is occupied by one of these pairs of letters in accordance with the following rules:

(1) The A series is started by placing the pair $[A + a]$ in any desired cell, say cell $(1,0)$.

(2) The remaining pairs in the A series are located by taking a series of construction steps, which we shall identify as the (C,R) step, consisting of C columns to the right (left when C is negative) and R rows up (down when R is negative) from the cell last filled. In our example, Case 1, C is $+2$ and R is $+1$, so the step is two columns over to the right and one row up. In the case of a fifth-order square this will place $[A + b]$ in cell $(3,1)$; $[A + c]$ in cell $(5,2)$ or, what is the equivalent, $(0,2)$; $[A + d]$ in cell $(2,3)$; and $[A + e]$ in cell $(4,4)$. This completes the A series. Note that if you attempted to take one more regular (C,R) step the next cell $(6,5)$ or, what is the equivalent, $(1,0)$ is already occupied by $[A + a]$.

(3) It is evident that a special move (which we shall refer to as a cross-step) is necessary to start the B series. This cross-step, which we shall identify as the $(C + c, R + r)$ cross-step, consists of a step of $(C + c)$ columns to the right (left if $C + c$ is negative) and $(R + r)$ rows up (down if $R + r$ is negative). In our example, Case 1, $C + c = +1$ and $R + r = -1$, so the cross-step is one column to the right and one row down. It follows that $[B + a]$ belongs in cell $(5,3)$ or, what is the equivalent, $(0,3)$. See Figure 36b for the appearance of the square at this point.

(4) After the cross-step has located the first pair of the B series, continue with the regular (C,R) step until the B series is completed. At this point, start the C series by again taking the cross-step $(C + c, R + r)$. Figure 36c shows the square at this point.

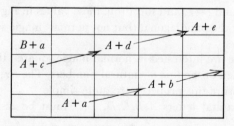

Fig. 36b.

		$B+b$		$A+e$
$B+a$		$A+d$		
$A+c$			$B+e$	
	$B+d$		$A+b$	$C+a$
	$A+a$			$B+c$

Fig. 36c.

Repeat the above process until the square is completed as shown in Figure 36d.

$C+d$	$E+c$	$B+b$	$D+a$	$A+e$
$B+a$	$D+e$	$A+d$	$C+c$	$E+b$
$A+c$	$C+b$	$E+a$	$B+e$	$D+d$
$E+e$	$B+d$	$D+c$	$A+b$	$C+a$
$D+b$	$A+a$	$C+e$	$E+d$	$B+c$

Fig. 36d.

Notice that in each row, column, main diagonal and broken diagonal the letters A, B, C, D and E appear once, and only once. Also note that the letters a, b, c, d and e appear once, and only once. In other words, if numerical values are assigned to these letters the resultant square will be pandiagonal regardless of the values selected. Suppose, for example, that we desired to construct a pandiagonal magic square whose bottom row consisted of π to eight decimals. If you let $A = 10$, $B = 40$, $C = 14$, $D = 0$, $E = 90$, $a = 4$, $b = 3$, $c = 25$, $d = 2$ and $e = 1$, and substitute these values in Figure 36d, you will generate the pandiagonal (but not normal) magic square shown in Figure 36e.

Usually we are interested in normal magic squares, that is, magic squares formed of the first n^2 natural numbers. The requirements to be met to ensure this property are quite simple:

(1) The capital letters $A, B, C, D, \ldots, N$ must be assigned values from:

(a) either the set $0, n, 2n, 3n, \ldots, (n-1)n$, or

(b) the set $1, 2, 3, 4, \ldots, n$.

16	115	43	4	11
44	1	12	39	93
35	17	94	41	2
91	42	25	13	18
3	14	15	92	65

Fig. 36e. $\pi = 3.14159265!!!$

(2) The lower-case letters $a, b, c, d, \ldots, n$ must be assigned values from:

(a) the set $1, 2, 3, 4, \ldots, n$ whenever the capital letters are assigned values from the set $0, n, 2n, 3n, \ldots, (n-1)n$, or

(b) the set $0, n, 2n, 3n, \ldots, (n-1)n$ whenever the capital letters are assigned values from the set $1, 2, 3, 4, \ldots, n$.

It is absolutely immaterial in what order you assign these values as long as you assign each one once, and only once. In all cases you will generate a normal pandiagonal magic square.

In the case of a fifth-order square these values become:

(1) the capital letters 0, 5, 10, 15 and 20 and the lower-case letters 1, 2, 3, 4 and 5, or

(2) the capital letters 1, 2, 3, 4 and 5 and the lower-case letters 0, 5, 10, 15 and 20.

There are, therefore, 10 ways we can select A (any one of either set). For each of these 10 ways there are 4 ways we can select B (any one of the four remaining values in the set from which A was selected). For each of these 40 ways there are 3 ways we can select C. For each of these 120 ways there are 2 ways we can select D and, finally, only one way left to select E. In a similar way we see that for each of these 240 ways there are $5 \times 4 \times 3 \times 2 \times 1 = 120$ ways of selecting values for the lower-case letters. This gives us a grand total of $240 \times 120 = 28,800 = 2(5!)^2$* normal pandiagonal magic squares that can be generated from the basic square shown in Figure 36d.

It should be noted, however, that these squares are not all different. Using Figure 1 as an example, we see that (as shown in Figure 37) there are eight ways any one square can be viewed. For purposes of enumeration, these eight squares are considered as being equivalent to just one square.

---

*Here (5!), or five factorial, is used in the normal mathematical sense to indicate $5 \times 4 \times 3 \times 2 \times 1$.

If we divide 28,800 by 8 to eliminate this duplication, we find that Figure 36d can be used to generate $3,600 = (5!)^2/4$ different pandiagonal fifth-order magic squares. This is all there are! Truly this square, from which all of the magic squares of this type can be generated, belongs in Franklin's collection of magical magic squares!

8	1	6
3	5	7
4	9	2

#1

4	3	8
9	5	1
2	7	6

#1 rotated 90°

2	9	4
7	5	3
6	1	8

#1 rotated 180°

6	7	2
1	5	9
8	3	4

#1 rotated 270°

4	9	2
3	5	7
8	1	6

2	7	6
9	5	1
4	3	8

6	1	8
7	5	3
2	9	4

8	3	4
1	5	9
6	7	2

Mirror images of squares immediately above

Fig. 37.

It is obvious that, by the proper selection of values any desired number can be located in any particular cell. If the additional property of symmetry is desired it is only necessary to so assign the values that the middle row is symmetrical. For example, let $A = 15$, $B = 20$, $C = 0$, $D = 5$, $E = 10$, $a = 3$, $b = 2$, $c = 5$, $d = 1$ and $e = 4$. Figure 38a is the result. In this form it is a very simple matter to check that the final square, Figure 38b, will be a normal, symmetrical, pandiagonal magic square as was predicted.

0 + 1	10 + 5	20 + 2	5 + 3	15 + 4
20 + 3	5 + 4	15 + 1	0 + 5	10 + 2
15 + 5	0 + 2	10 + 3	20 + 4	5 + 1
10 + 4	20 + 1	5 + 5	15 + 2	0 + 3
5 + 2	15 + 3	0 + 4	10 + 1	20 + 5

Fig. 38a.

1	15	22	8	19
23	9	16	5	12
20	2	13	24	6
14	21	10	17	3
7	18	4	11	25

Fig. 38b.   A fifth-order, symmetrical, pandiagonal magic
square.

Let us now apply the same approach to an eighth-order square. If
$[A + a]$ is again placed in cell $(1,0)$ the pair $[A + b]$ will fall in the
cell which is two columns to the right and one row up, or in cell
$(3,1)$; the pair $[A + c]$ in the cell $(5,2)$; the pair $[A + d]$ in cell $(7,3)$;
the pair $[A + e]$ in cell $(9,4)$ or, what is the equivalent, $(1,4)$; the
pair $[A + f]$ in cell $(3,5)$; the pair $[A + g]$ in cell $(5,6)$; and the final
pair in the A series $[A + h]$ in cell $(7,7)$. Using the same cross-step as
before, one column to the right and one row down, will place the
first pair in the B series $[B + a]$ in cell $(8,6)$ or, what is the
equivalent, $(0,6)$. Figure 39a shows the appearance of the intermediate square at this point.

							$A + h$
$B + a$					$A + g$		
			$A + f$				
	$A + e$						
							$A + d$
					$A + c$		
			$A + b$				
	$A + a$						

Fig. 39a.

Continuing this process until the intermediate square is finished
will generate Figure 39b. Examination of this figure will show that
in the case of each row, column, main diagonal and broken dia-

gonal the letters A, B, C, D, E, F, G and H, as well as a, b, c, d, e, f, g and h appear once, and only once, except in the case of the capital letters in the columns and the lower-case letters in the rows. However, these two exceptions will have the correct totals if we select values so that:

$$A + C + E + G = B + D + F + H,$$

and

$$a + c + e + g = b + d + f + h.$$

$D+f$	$G+d$	$B+b$	$E+h$	$H+f$	$C+d$	$F+b$	$A+h$
$B+a$	$E+g$	$H+e$	$C+c$	$F+a$	$A+g$	$D+e$	$G+c$
$H+d$	$C+b$	$F+h$	$A+f$	$D+d$	$G+b$	$B+h$	$E+f$
$F+g$	$A+e$	$D+c$	$G+a$	$B+g$	$E+e$	$H+c$	$C+a$
$D+b$	$G+h$	$B+f$	$E+d$	$H+b$	$C+h$	$F+f$	$A+d$
$B+e$	$E+c$	$H+a$	$C+g$	$F+e$	$A+c$	$D+a$	$G+g$
$H+h$	$C+f$	$F+d$	$A+b$	$D+h$	$G+f$	$B+d$	$E+b$
$F+c$	$A+a$	$D+g$	$G+e$	$B+c$	$E+a$	$H+g$	$C+e$

Fig. 39b.

Also note that, just as in the case of the fifth-order square, all the letters are symmetrically located with respect to the center of the square. It follows that if we assign the values 0, 8, 16, 24, 32, 40, 48 and 56 to the capital letters and 1, 2, 3, 4, 5, 6, 7 and 8 to the lower-case letters (or the reverse) so that symmetrically located pairs are complementary (that is, add up to 56 or 9 as the case may be) and the above conditions are met, the result will be a normal, symmetrical, pandiagonal magic square of the eighth order. There are many different ways that this can be done. For example, let $A = 0$ (since F is symmetrically located with respect to A, it follows that we have to make $F = 56 - 0 = 56$); $C = 24$ (and, therefore, $D = 32$); $E = 40$ (and, therefore, $B = 16$); $G = 48$ (and, therefore, $H = 8$); $a = 1$ (and $b = 9 - 1 = 8$); $c = 4$ (and $h = 5$); $e = 6$ (and $f = 3$); and finally $g = 7$ (and $d = 2$).

Substituting these values, which meet the above conditions, in Figure 39b will generate Figure 40a. Examination of this square, and that of Figure 40b (which was derived from it by adding the pairs of numbers in the same cell together), will show that they are —as predicted—eighth-order, symmetrical, pandiagonal magic squares.

32 + 3	48 + 2	16 + 8	40 + 5	8 + 3	24 + 2	56 + 8	0 + 5
16 + 1	40 + 7	8 + 6	24 + 4	56 + 1	0 + 7	32 + 6	48 + 4
8 + 2	24 + 8	56 + 5	0 + 3	32 + 2	48 + 8	16 + 5	40 + 3
56 + 7	0 + 6	32 + 4	48 + 1	16 + 7	40 + 6	8 + 4	24 + 1
32 + 8	48 + 5	16 + 3	40 + 2	8 + 8	24 + 5	56 + 3	0 + 2
16 + 6	40 + 4	8 + 1	24 + 7	56 + 6	0 + 4	32 + 1	48 + 7
8 + 5	24 + 3	56 + 2	0 + 8	32 + 5	48 + 3	16 + 2	40 + 8
56 + 4	0 + 1	32 + 7	48 + 6	16 + 4	40 + 1	8 + 7	24 + 6

Fig. 40a.

35	50	24	45	11	26	64	5
17	47	14	28	57	7	38	52
10	32	61	3	34	56	21	43
63	6	36	49	23	46	12	25
40	53	19	42	16	29	59	2
22	44	9	31	62	4	33	55
13	27	58	8	37	51	18	48
60	1	39	54	20	41	15	30

Fig. 40b.   An eighth-order, symmetrical, pandiagonal magic square.

We now apply the same approach to a tenth-order square. Figure 41 is the result. On the face of it this square is similar to the eighth-order square that we just completed. The rows, columns, main diagonals and broken diagonals all contain the letters once, and only once, except that—as before—the capital letters in the columns and the lower-case letters in the rows are duplicated. For these two exceptions to be correct it is necessary for the values to be selected so that:

$$A + C + E + G + I = B + D + F + H + J,$$

and

$$a + c + e + g + i = b + d + f + h + j.$$

$H+d$	$E+h$	$B+b$	$I+f$	$F+j$	$C+d$	$J+h$	$G+b$	$D+f$	$A+j$
$B+a$	$I+e$	$F+i$	$C+c$	$J+g$	$G+a$	$D+e$	$A+i$	$H+c$	$E+g$
$F+h$	$C+b$	$J+f$	$G+j$	$D+d$	$A+h$	$H+b$	$E+f$	$B+j$	$I+d$
$J+e$	$G+i$	$D+c$	$A+g$	$H+a$	$E+e$	$B+i$	$I+c$	$F+g$	$C+a$
$D+b$	$A+f$	$H+j$	$E+d$	$B+h$	$I+b$	$F+f$	$C+j$	$J+d$	$G+h$
$H+i$	$E+c$	$B+g$	$I+a$	$F+e$	$C+i$	$J+c$	$G+g$	$D+a$	$A+e$
$B+f$	$I+j$	$F+d$	$C+h$	$J+b$	$G+f$	$D+j$	$A+d$	$H+h$	$E+b$
$F+c$	$C+g$	$J+a$	$G+e$	$D+i$	$A+c$	$H+g$	$E+a$	$B+e$	$I+i$
$J+j$	$G+d$	$D+h$	$A+b$	$H+f$	$E+j$	$B+d$	$I+h$	$F+b$	$C+f$
$D+g$	$A+a$	$H+e$	$E+i$	$B+c$	$I+g$	$F+a$	$C+e$	$J+i$	$G+c$

Fig. 41.

But the sum of $0 + 10 + 20 + 30 + 40 + 50 + 60 + 70 + 80 + 90 = 450$ and the sum of $1 + 2 + 3 + 4 + 5 + 6 + 7 + 8 + 9 + 10 = 55$ and it is not possible to split either of these two groups into two equal amounts. This will be true for all singly-even squares.

While the above conditions cannot be met, it is possible to assign values so that the totals of the even-numbered columns will be too large by v and the totals of the odd-numbered columns will be too small by v. Similarly, the totals of the even-numbered rows can be made too small by w and the totals of the odd-numbered rows too large by w. In other words, it is possible to assign values so that the following equations will be correct:

$$2(A + C + E + G + I) + v = 2(B + D + F + H + J) - v,$$

and

$$2(a + c + e + g + i) + w = 2(b + d + f + h + j) - w.$$

Now, if we let $J - E = v$ we can interchange $[E + h]$ and $[J + h]$ in the top, or ninth, row without changing the total of the row. We will, however, have increased the total of column 1 by v and decreased the total of column 6 by v and thus made these two columns correct. A similar interchange between $[J+g]$ and $[E+g]$ in row 8, between $[J+j]$ and $[E+j]$ in row 1, and $[J+i]$ and $[E+i]$ in row 0, will make all columns correct except 2 and 7. If we were to interchange either $[J+f]$ and $[E+f]$ in row 7 or $[J+a]$ and $[E+a]$ in row 2, we would correct these columns but we would, at the same time, be introducing an error in the main diagonals. This can be avoided by making the interchange elsewhere, rather than on a main diagonal, provided we assign the proper set of values to another pair of capital letters. For example, let $B-G=v$ and interchange $[B+b]$ and $[G+b]$ in row 9. The overall result of all of these changes will be to make the columns correct without affecting the totals of the rows or the main diagonals.

We can correct the rows in a similar manner. One word of caution here—do not interchange the capital letters and the lower case letters in the same cell. Let $h-c=f-a=w$; then the following interchanges will make the totals of the rows correct:

$[F+c]$ and $[F+h]$ in column 0, $[B+f]$ and $[B+a]$ in column 0,

$[A+a]$ and $[A+f]$ in column 1, $[D+a]$ and $[D+f]$ in column 8,

and $[C+f]$ and $[C+a]$ in column 9.

Note that by making these interchanges (which are necessary to make the square magic) you destroy the property of symmetry and make some of the broken diagonals incorrect. There is no way to avoid this. As mentioned earlier, you cannot construct a symmetrical or pandiagonal singly-even-order magic square.

Meeting these limitations on the values to be selected is not very difficult. For example, let $A=30$, $B=90$, $C=70$, $D=10$, $E=40$, $F=20$, $G=80$, $H=60$, $I=0$, $J=50$, $a=9$, $b=3$, $c=1$, $d=7$, $e=4$, $f=10$, $g=5$, $h=2$, $i=8$ and $j=6$. Then:

$$2(A+C+E+G+I)+10=2(30+70+40+80+0)+10=450,$$

$$2(B+D+F+H+J)-10=2(90+10+20+60+50)-10=450,$$

$$J-E=B-G=10, \qquad \text{and}$$

$$2(a+c+e+g+i)+1=2(9+1+4+5+8)+1=55,$$

$$2(b+d+f+h+j)-1=2(3+7+10+2+6)-1=55,$$

$$h-c=f-a=1, \qquad \text{as required.}$$

Figure 42 is the result of making these interchanges and substitutions.

60 + 7	50 + 2	80 + 3	0 + 10	20 + 6	70 + 7	40 + 2	90 + 3	10 + 9	30 + 6
90 + 10	0 + 4	20 + 8	70 + 1	40 + 5	80 + 9	10 + 4	30 + 8	60 + 1	50 + 5
20 + 1	70 + 3	50 + 10	80 + 6	10 + 7	30 + 2	60 + 3	40 + 10	90 + 6	0 + 7
50 + 4	80 + 8	10 + 1	30 + 5	60 + 9	40 + 4	90 + 8	0 + 1	20 + 5	70 + 10
10 + 3	30 + 9	60 + 6	40 + 7	90 + 2	0 + 3	20 + 10	70 + 6	50 + 7	80 + 2
60 + 8	40 + 1	90 + 5	0 + 9	20 + 4	70 + 8	50 + 1	80 + 5	10 + 10	30 + 4
90 + 9	0 + 6	20 + 7	70 + 2	50 + 3	80 + 10	10 + 6	30 + 7	60 + 2	40 + 3
20 + 2	70 + 5	50 + 9	80 + 4	10 + 8	30 + 1	60 + 5	40 + 9	90 + 4	0 + 8
40 + 6	80 + 7	10 + 2	30 + 3	60 + 10	50 + 6	90 + 7	0 + 2	20 + 3	70 + 9
10 + 5	30 + 10	60 + 4	50 + 8	90 + 1	0 + 5	20 + 9	70 + 4	40 + 8	80 + 1

Fig. 42.

Figure 43, which is formed from Figure 42 by adding together the two numbers in each cell, is thus magic but not symmetrical or pandiagonal.

67	52	83	10	26	77	42	93	19	36
100	4	28	71	45	89	14	38	61	55
21	73	60	86	17	32	63	50	96	7
54	88	11	35	69	44	98	1	25	80
13	39	66	47	92	3	30	76	57	82
68	41	95	9	24	78	51	85	20	34
99	6	27	72	53	90	16	37	62	43
22	75	59	84	18	31	65	49	94	8
46	87	12	33	70	56	97	2	23	79
15	40	64	58	91	5	29	74	48	81

Fig. 43.   A tenth-order magic square.

The making of a 6-by-6 magic square by this method involves a little more care, but actually it is not much more difficult. We cannot, however, use the values of C, R, c and r given in Case 1 because in that case $(Cr - cR) = 3$ is not prime to 6. Suppose we select $C = 2$, $R = 1$, $c = 1$ and $r = -2$. Then $(Cr - cR) = -5$, which is prime to 6. The regular step will be two columns to the right and one row up (as in Case 1) and the cross-step will be three columns to the right and one row down. Using these values and following our normal procedure, we will generate Figure 44a.

Making the shifts to correct the totals of the columns and rows will present an additional problem. You will find that you cannot avoid making a shift involving the main diagonals. This will, of course, introduce an error into the total of the diagonal involved in the shift. It is, therefore, necessary to make another shift involving the same diagonal which cancels the error introduced by the first shift. Figure 44a demonstrates that this really presents no serious problem. Examination will show that if we can assign values to the letters so as to meet the following requirements we will be able to make the necessary transfers:

$F+d$	$E+b$	$D+f$	$C+d$	$B+b$	$A+f$
$D+e$	$C+c$	$B+a$	$A+e$	$F+c$	$E+a$
$B+f$	$A+d$	$F+b$	$E+f$	$D+d$	$C+b$
$F+a$	$E+e$	$D+c$	$C+a$	$B+e$	$A+c$
$D+b$	$C+f$	$B+d$	$A+b$	$F+f$	$E+d$
$B+c$	$A+a$	$F+e$	$E+c$	$D+a$	$C+e$

Fig. 44a.

$$2(A + C + E) + v = 2(B + D + F) - v,$$
$$B - E = D - A = v,$$
$$F + C = (A + B + C + D + E + F)/3,$$
$$2(a + c + e) + w = 2(b + d + f) - w,$$
$$b - e = d - a = w,$$
$$f + c = (a + b + c + d + e + f)/3.$$

These conditions can be met in various ways. One way is to let:

$A = 24$, $B = 6$, $C = 18$, $D = 30$, $E = 0$, $F = 12$, $v = 6$, and
$a = 5$, $b = 2$, $c = 4$, $d = 6$, $e = 1$, $f = 3$, and $w = 1$.

Making the transfers indicated in Figure 44a, and inserting the values assigned above, gives us Figures 44b and 44c.

12+5	6+2	30+3	18+6	0+2	24+3
30+2	18+4	6+5	24+1	12+4	0+5
0+3	24+6	12+2	6+3	30+6	18+1
12+6	0+1	24+4	18+5	6+1	30+4
30+1	18+3	6+6	24+2	12+3	0+6
6+4	24+5	12+1	0+4	30+5	18+2

Fig. 44b.

17	8	33	24	2	27
32	22	11	25	16	5
3	30	14	9	36	19
18	1	28	23	7	34
31	21	12	26	15	6
10	29	13	4	35	20

Fig. 44c.   A sixth-order magic square.

Let us now turn to Case 2. Here the regular (C, R) step becomes a $(1, 1)$ step; that is, one column to the right and one row up. Also, the $(C + c, R + r)$ cross-step becomes a $(0, 1)$ cross-step; that is, one row down in the same column. If the A series is started in the cell $(2, 4)$ the square shown in Figure 45a will be generated.

$D+b$	$E+d$	$A+a$	$B+c$	$C+e$
$E+c$	$A+e$	$B+b$	$C+d$	$D+a$
$A+d$	$B+a$	$C+c$	$D+e$	$E+b$
$B+e$	$C+b$	$D+d$	$E+a$	$A+c$
$C+a$	$D+c$	$E+e$	$A+b$	$B+d$

Fig. 45a.

Notice that in Figure 45a, except for the capital letters in the upwards diagonals, each row, column, main diagonal and broken diagonal contains each letter once, and only once. In other words, if numerical values are assigned to these letters the resultant square will be at least semimagic. Note further that, no matter how the values are assigned, it is not possible to make the square pandiagonal. It is possible, however, to make it magic if the capital letter C is assigned a value equal to the average of A, B, C, D and E. Also, if we meet the conditions previously given, it is possible to make the square both normal and symmetrical. Suppose we let $A = 0$ (and E, which is symmetrically located to A, equal 20), $B = 5$ (and D, which is symmetrically located to B, equal 15), $C = 10$ (the average of 0, 5, 10, 15 and 20), $a = 1$ (and e, which is symmetrically located to a, equal 5), $b = 2$ (and d, which is symmetrically located to b, equal 4) and $c = 3$ (the average of 1, 2, 3, 4 and 5). Substituting these values in Figure 45a gives us Figures 45b and 45c.

15 + 2	20 + 4	0 + 1	5 + 3	10 + 5
20 + 3	0 + 5	5 + 2	10 + 4	15 + 1
0 + 4	5 + 1	10 + 3	15 + 5	20 + 2
5 + 5	10 + 2	15 + 4	20 + 1	0 + 3
10 + 1	15 + 3	20 + 5	0 + 2	5 + 4

Fig. 45b.

17	24	1	8	15
23	5	7	14	16
4	6	13	20	22
10	12	19	21	3
11	18	25	2	9

Fig. 45c.   A fifth-order symmetrical magic square.

Notice that Figure 45c is identical with Figure 16d. Thus we see that, by carefully selecting the values assigned to the letters, Case 2 reduces to the De la Loubère rule. In other words, the De la

Loubère rule is merely a special case under our method. The same is true of the Méziriac rule.

To summarize the description of our new cyclical method as far as we have gone:

(1) It has been demonstrated that, by the use of an intermediate square composed of letters, it is possible to extend the cyclical construction process to even-order, as well as odd-order, magic squares.

(2) The numerical values selected for use in the previous examples were chosen more or less at random. There are many other combinations which would have worked equally well. As demonstrated by the fact that every one of the 3,600 possible fifth-order pandiagonal magic squares can be generated by assigning the proper values to a single basic intermediate square, this freedom of choice adds greatly to the flexibility of the method.

(3) It now remains to:

(a) define the $\{p,q\}$ magic series, and

(b) show how the magical properties of the final square depend upon the choice of the four constants C, R, c and r.

10

SPECIAL MAGICAL PROPERTIES

Let us look again at Figure 17d, repeated on page 61 for convenience, and note that the seven circled numbers 22, 41, 4, 16, 35, 47 and 10 (which were determined by taking 22 as an arbitrary starting point and then using a (3, 1) step to select the other six) total 175, which is the magic constant for a seventh-order square. Note also that you may take any one of the 49 cells as a starting point and, by using a (3, 1) step six times, obtain a series of seven numbers that will add up to 175 in each case. In other words, this square is magic for the {3, 1} series even though it is not pandiagonal. This concept of a magic series may be stated formally as follows:

(1) Define a *specific* $\{p,q\}$ *series* as being a series of n numbers determined by taking some cell as a starting point and then taking $(n-1)$ steps of p columns to the right (left when p is negative) and q rows up (down when q is negative) and then:

(2) Define a $\{p,q\}$ *series* as being all possible specific $\{p,q\}$ series.

As would be expected, a $\{p,q\}$ series is considered magic when each of the specific $\{p,q\}$ series which compose it has a total equal to the magic constant of the square. It is evident that the $\{1,1\}$ series is, by definition, equivalent to the upward main and broken diagonals and, similarly, the $\{1,-1\}$ series is equivalent to the downward main and broken diagonals. It follows that if the $\{1,1\}$ and the $\{1,-1\}$ series are both magic, along with the rows and columns, the square will be a pandiagonal magic square.

Just as the final properties of the square depend upon the numerical values assigned to the capital and lower-case letters, the manner in which these letters fall in any row, column, diagonal or series (of the intermediate square) depends upon the values assigned to C, R, c and r. The following table lists for future reference the controlling characteristic (which will be C, R, c, r or some combination of these quantities) in each instance:

TABLE OF CONTROLLING CHARACTERISTICS

Magic Series in Question	Controlling Characteristic for Capital Letters	Controlling Characteristic for Lower-case Letters
Columns	C	c
Rows	R	r
$\{1,1\}$ series or, what is the same thing, the upward diagonals	$(C-R)$	$(c-r)$
$\{1,-1\}$ series or, what is the same thing, the downward diagonals	$(C+R)^*$	$(c+r)^*$
Any $\{p,q\}$ series	$(qC-pR)$	$(qc-pr)$

*Actually $(-C-R)$ and $(-c-r)$ but, since the sign of the controlling characteristic makes no difference, we refer to it as $(C+R)$ and $(c+r)$.

In the following discussion we will cover capital letters only. What is said, however, is equally applicable to the lower-case letters. The number of different capital letters in any series (including the rows and columns, which are after all merely special cases of

30	39	48	1	⑩	19	28
38	㊼	7	9	18	27	29
46	6	8	17	26	㉟	37
5	14	⑯	25	34	36	45
13	15	24	33	42	44	④
21	23	32	㊶	43	3	12
㉒	31	40	49	2	11	20

Fig. 17d.

the more general series, the rows being equivalent to the series $\{p,0\}$ and the columns to the series $\{0,q\}$) will depend upon the greatest common divisor—hereafter referred to as g.c.d.—of the controlling characteristic and n. For example, in the case of the capital letters in the columns it will be the g.c.d. of C and n; in the case of the capital letters in the upward diagonals it will be the g.c.d. of $(C-R)$ and n; and so forth. If the g.c.d. is 1—that is, if the two numbers are prime to each other—all n of the capital letters in the series controlled by that particular characteristic will be different. If the g.c.d. is 2, there will be $n/2$ different capital letters, each appearing twice. If the g.c.d. is 3, there will be $n/3$ different capital letters, each appearing three times. And so forth up to the case where the g.c.d. is n (that is, the controlling characteristic is zero or some multiple of n). In this case the same capital letter will appear in all terms in the series.

The effect of the above relations, and those previously discussed in Chapter 9, may be summarized as follows:

CONDITION 1: *All squares.* In order for all pairs of letters to fall into separate distinct cells it is necessary that $(Cr-cR)$ be prime to n. In order that no row or column contain only one capital (or lower-case) letter it is necessary that none of the controlling characteristics C, R, c and r equal zero.

CONDITION 2: *Odd-order squares.* Since C, R, c and r cannot be zero, the rows, columns and main diagonals will either be automatically magic (each series will consist of n different letters, each appearing once and only once) or can be made magic by proper selection of numerical values for the letters.

CONDITION 3: *Even-order squares*. If the number of different capital (lower-case) letters in any given series is even—that is, if n divided by the g.c.d. of the controlling characteristic and n is even—it is possible to make the series magic by the proper selection of numerical values for the letters. On the other hand, when the number of different capital (lower-case) letters in any given series is odd, in addition to assigning the numerical values carefully it will be necessary to interchange certain pairs of letters before the square can be made magic. In the case of singly-even squares, however, if C and r are both odd and R and c are both even (or the reverse) and if, in addition, all four quantities are prime to $n/2$, it will always be possible to select numerical values for the letters which will (after the necessary interchanges to make the rows and columns magic) make the final square magic.

Let us return for a moment to our previous examples, Case 1 and Case 2. The values of the controlling characteristics for the diagonals are as follows:

Series	Controlling Characteristic	Case 1	Case 2
Upward Diagonals:			
Capital Letters	$(C-R)$	+1	0
Lower-case Letters	$(c-r)$	+1	+1
Downward Diagonals:			
Capital Letters	$(C+R)$	+3	+2
Lower-case Letters	$(c+r)$	−3	−3

In Case 1 all four of these values are prime to 5, 8 and 10. It follows that we could have predicted in advance that each capital and lower-case letter would appear once, and only once, in the upward and downward diagonals in Figures 36d, 39b and 41 (before the interchanges were made). Note also that in Case 1 both C and r were even and both R and c were odd and that for the singly-even square, where n was 10, all four were prime to $n/2=5$, hence we knew in advance that the necessary interchanges could be made.

From the fact that $(C-R)$ equals 0 in Case 2 we could have predicted in advance of actual construction of the intermediate square that the capital letters in the upward diagonals in Figure 45a would all be alike and that it would not be possible to assign numerical values to the letters so as to make the square pandiagonal. The best that can be done is to so assign the letters on the main upward diagonal, C, that the square will be magic.

Suppose we desire to construct an odd-order magic square which is both symmetrical and pandiagonal and which has the additional property that all knight's-move series are correct. A knight's move is, of course, a move of two cells across and one cell up or down, or a move of one cell across and two cells up or down. The eight possible knight's moves generate the eight series $\{2,1\}$, $\{2,-1\}$, $\{-2,1\}$, $\{-2,-1\}$, $\{1,2\}$, $\{1,-2\}$, $\{-1,2\}$, $\{-1,-2\}$. These may be represented as $\{\pm2,\pm1\}$ and $\{\pm1,\pm2\}$.

For reasons which will appear later let $C=3$, $R=1$, $c=1$ and $r=3$. This gives $(Cr-cR)=(9-1)=8$ and the following values to the controlling characteristics:

$$(C-R) = 2 \qquad\qquad (c-r) = -2$$

$$(C+R) = 4 \qquad\qquad (c+r) = 4$$

$$(\pm2C\pm R) = \pm5 \text{ or } \pm7 \qquad (\pm2c+r) = \pm1 \text{ or } \pm5$$

$$(\pm C\pm2R) = \pm1 \text{ or } \pm5 \qquad (\pm c\pm2r) = \pm5 \text{ or } \pm7$$

We observe that the smallest value of n prime to all these controlling characteristics is 9. Accordingly we can predict that a ninth-order square made with step $(3,1)$ and cross-step $(4,4)$ will have each row, column, upward and downward diagonal, plus the $\{\pm2,\pm1\}$ and $\{\pm1,\pm2\}$ series consist of each capital and each lower-case letter once, and only once, except that the capital letters in the columns and the lower-case letters in the rows will consist of $9/3=3$ different letters, each appearing three times.

Figure 46, which is such a square, shows that it does have the properties predicted. In order for this square to have the desired magical properties it is only necessary that:

$$A+D+G=B+E+H=C+F+I=108,$$

$$a+d+g=b+e+h=c+f+i=15,$$

and that symmetrically located letters be assigned complementary values.

$D+i$	$E+f$	$F+c$	$G+i$	$H+f$	$I+c$	$A+i$	$B+f$	$C+c$
$G+h$	$H+e$	$I+b$	$A+h$	$B+e$	$C+b$	$D+h$	$E+e$	$F+b$
$A+g$	$B+d$	$C+a$	$D+g$	$E+d$	$F+a$	$G+g$	$H+d$	$I+a$
$D+f$	$E+c$	$F+i$	$G+f$	$H+c$	$I+i$	$A+f$	$B+c$	$C+i$
$G+e$	$H+b$	$I+h$	$A+e$	$B+b$	$C+h$	$D+e$	$E+b$	$F+h$
$A+d$	$B+a$	$C+g$	$D+d$	$E+a$	$F+g$	$G+d$	$H+a$	$I+g$
$D+c$	$E+i$	$F+f$	$G+c$	$H+i$	$I+f$	$A+c$	$B+i$	$C+f$
$G+b$	$H+h$	$I+e$	$A+b$	$B+h$	$C+e$	$D+b$	$E+h$	$F+e$
$A+a$	$B+g$	$C+d$	$D+a$	$E+g$	$F+d$	$G+a$	$H+g$	$I+d$

Fig. 46.

There are various ways of meeting the above requirements, one of which is as follows: $A=72$ (making $C=0$), $D=9$ (making $I=63$), $E=54$ (making $H=18$), $F=45$ (making $G=27$), $B=36$, $a=9$ (making $c=1$), $d=2$ (making $i=8$), $e=7$ (making $h=3$), $f=6$ (making $g=4$) and $b=5$. Substituting these values in Figure 46 gives Figure 47a.

From this square the generation of Figure 47b, the final square, is simple.

Examination will show that not only does this square have all the magically magic properties predicted for it, it also has the unusual property that the nine numbers in any 3-by-3 subsquare will also add up to the magic constant 369. We have met this property before in the Franklin square, and in Figure 14, where the total of any four numbers forming a 2-by-2 subsquare equaled 130.

This property can also be predicted in advance. Before stating the conditions which must be met it is, however, desirable to broaden our definition of magic.

Since the n^2 numbers in a normal magic square total $(n^2)(n^2+1)/2$, their average value will be $(n^2+1)/2$. Let us, therefore, now define any series as being magic when—regardless of how the series was determined and regardless of the number of terms in it—the total of the k numbers forming the series equals $k(n^2+1)/2$. As would be expected, this criterion reduces to the normal magic constant, $n(n^2+1)/2$, when the number of terms in the series, k, equals the order of the square, n.

CONDITION 4: *Odd and doubly-even-order squares.* If r is made equal to $\pm C$, c equal to $\pm R$, y equal to n divided by the g.c.d. of n and r,

9+8	54+6	45+1	27+8	18+6	63+1	72+8	36+6	0+1
27+3	18+7	63+5	72+3	36+7	0+5	9+3	54+7	45+5
72+4	36+2	0+9	9+4	54+2	45+9	27+4	18+2	63+9
9+6	54+1	45+8	27+6	18+1	63+8	72+6	36+1	0+8
27+7	18+5	63+3	72+7	36+5	0+3	9+7	54+5	45+3
72+2	36+9	0+4	9+2	54+9	45+4	27+2	18+9	63+4
9+1	54+8	45+6	27+1	18+8	63+6	72+1	36+8	0+6
27+5	18+3	63+7	72+5	36+3	0+7	9+5	54+3	45+7
72+9	36+4	0+2	9+9	54+4	45+2	27+9	18+4	63+2

Fig. 47a.

17	60	46	35	24	64	80	42	1
30	25	68	75	43	5	12	61	50
76	38	9	13	56	54	31	20	72
15	55	53	33	19	71	78	37	8
34	23	66	79	41	3	16	59	48
74	45	4	11	63	49	29	27	67
10	62	51	28	26	69	73	44	6
32	21	70	77	39	7	14	57	52
81	40	2	18	58	47	36	22	65

Fig. 47b.   A symmetrical, pandiagonal, ninth-order magic
square with added magical properties.

and z equal to n divided by the g.c.d. of n and c, than any selection
of numerical values for the letters which will make the rows and
columns magic *without interchanging any pairs* will also result in all
y-by-y and z-by-z subsquares being magic.

We are now in a position to state the condition which will ensure
that Franklin's bent diagonals, or his half-rows and half-columns,
will be magic.

CONDITION 5: *Doubly-even-order squares.* If C equals r and is prime
to n and if c equals R equals $n/2$, the square will be pandiagonal,
with the exception that there will be only two lower-case letters in
each column and only two capital letters in each row. (Or, if you

prefer, you could make c equal to R prime to n and C equal to r equal to $n/2$.) By proper assignment of numerical values to the letters it will be possible to make the rows and columns magic, retain the pandiagonal property, make the 2-by-2 subsquares magic (this will follow automatically from the selection of values to make the columns and rows magic) and—for n greater than 4—to make either the half-rows and half-columns, or the bent diagonals, magic.

As a final example of this method let us attempt to construct an eighth-order magic square which will be symmetrical, pandiagonal, have magical 2-by-2 subsquares and bent diagonals, and will have as many $\{p,q\}$ series magic as possible.

Let $C = r = 5$ and $R = c = 4$. Remembering that $(jn + x) \equiv x \bmod n$ we have:

$$(Cr - cR) \equiv 25 - 16 \equiv 9 \equiv 1 \bmod 8,$$
$$(C \pm R) \equiv 5 \pm 4 \equiv 9 \text{ or } 1 \equiv 1 \bmod 8, \text{ and}$$
$$(c \pm r) \equiv 4 \pm 5 \equiv 9 \text{ or } -1 \equiv \pm 1 \bmod 8.$$

It follows that our square will be pandiagonal when the rows and columns are magic. Also, since $C = r = 5$ (is prime to 8) and $R = c = 4$ (equals 8/2), z will equal 2 and all 2-by-2 subsquares will be magic (when numerical values are selected which will make the rows and columns correct). Checking now on all possible $\{p,q\}$ series we have:

Series	Controlling Characteristics	
$\{\pm1,\pm2\}$	$(\pm2C\pm R) \equiv \pm14 \text{ or } \pm 6 \equiv \pm2$	$(\pm2c\pm r) \equiv \pm13 \text{ or } \pm 3 \equiv \pm3$
$\{\pm2,\pm1\}$	$(\pm C\pm2R) \equiv \pm13 \text{ or } \pm 3 \equiv \pm3$	$(\pm c\pm2r) \equiv \pm14 \text{ or } \pm 6 \equiv \pm2$
$\{\pm2,\pm2\}$	$(\pm2C\pm2R) \equiv \pm18 \text{ or } \pm 2 \equiv \pm2$	$(\pm2c\pm2r) \equiv \pm18 \text{ or } \pm 2 \equiv \pm2$
$\{\pm1,\pm3\}$	$(\pm3C\pm R) \equiv \pm19 \text{ or } \pm11 \equiv \pm3$	$(\pm3c\pm r) \equiv \pm17 \text{ or } \pm 7 \equiv \pm1$
$\{\pm3,\pm1\}$	$(\pm C\pm3R) \equiv \pm17 \text{ or } \pm 7 \equiv \pm1$	$(\pm c\pm3r) \equiv \pm19 \text{ or } \pm11 \equiv \pm3$
$\{\pm2,\pm3\}$	$(\pm3C\pm2R) \equiv \pm23 \text{ or } \pm 7 \equiv \pm1$	$(\pm3c\pm2r) \equiv \pm22 \text{ or } \pm 2 \equiv \pm2$
$\{\pm3,\pm2\}$	$(\pm2C\pm3R) \equiv \pm22 \text{ or } \pm 2 \equiv \pm2$	$(\pm2c\pm3r) \equiv \pm23 \text{ or } \pm 7 \equiv \pm1$
$\{\pm3,\pm3\}$	$(\pm3C\pm3R) \equiv \pm27 \text{ or } \pm 3 \equiv \pm3$	$(\pm3c\pm3r) \equiv \pm27 \text{ or } \pm 3 \equiv \pm3$
$\{\pm1,\pm4\}$	$(\pm4C\pm R) \equiv \pm24 \text{ or } \pm16 \equiv 0$	$(\pm4c\pm r) \equiv \pm21 \text{ or } \pm11 \equiv \pm3$
$\{\pm4,\pm1\}$	$(\pm C\pm4R) \equiv \pm21 \text{ or } \pm11 \equiv \pm3$	$(\pm c\pm4r) \equiv \pm24 \text{ or } \pm16 \equiv 0$
$\{\pm2,\pm4\}$	$(\pm4C\pm2R) \equiv \pm28 \text{ or } \pm12 \equiv \pm4$	$(\pm4c\pm2r) \equiv \pm26 \text{ or } \pm 6 \equiv \pm2$
$\{\pm4,\pm2\}$	$(\pm2C\pm4R) \equiv \pm26 \text{ or } \pm 6 \equiv \pm2$	$(\pm2c\pm4r) \equiv \pm28 \text{ or } \pm12 \equiv \pm4$
$\{\pm3,\pm4\}$	$(\pm4C\pm3R) \equiv \pm32 \text{ or } \pm 8 \equiv 0$	$(\pm4c\pm3r) \equiv \pm31 \text{ or } \pm 1 \equiv \pm1$
$\{\pm4,\pm3\}$	$(\pm3C\pm4R) \equiv \pm31 \text{ or } \pm 1 \equiv \pm1$	$(\pm3c\pm4r) \equiv \pm32 \text{ or } \pm 8 \equiv \pm0$
$\{\pm4,\pm4\}$	$(\pm4C\pm4R) \equiv \pm36 \text{ or } \pm 4 \equiv \pm4$	$(\pm4c\pm4r) \equiv \pm36 \text{ or } \pm 4 \equiv \pm4$

It is thus seen that all of the 49 possible $\{p,q\}$ series can be made magic by the proper selection of numerical values for the letters except the following eight series (remembering that $+4 \equiv -4 \bmod 8$): $\{4, \pm 1\}$, $\{4, \pm 3\}$, $\{\pm 1, 4\}$, $\{\pm 3, 4\}$. Figure 48 is the resultant square.

$D+e$	$H+b$	$D+g$	$H+d$	$D+a$	$H+f$	$D+c$	$H+h$
$G+a$	$C+f$	$G+c$	$C+h$	$G+e$	$C+b$	$G+g$	$C+d$
$B+e$	$F+b$	$B+g$	$F+d$	$B+a$	$F+f$	$B+c$	$F+h$
$E+a$	$A+f$	$E+c$	$A+h$	$E+e$	$A+b$	$E+g$	$A+d$
$H+e$	$D+b$	$H+g$	$D+d$	$H+a$	$D+f$	$H+c$	$D+h$
$C+a$	$G+f$	$C+c$	$G+h$	$C+e$	$G+b$	$C+g$	$G+d$
$F+e$	$B+b$	$F+g$	$B+d$	$F+a$	$B+f$	$F+c$	$B+h$
$A+a$	$E+f$	$A+c$	$E+h$	$A+e$	$E+b$	$A+g$	$E+d$

Fig. 48.

Examination of this square will show that if:

$$A + B + C + D = E + F + G + H = 112,$$
$$A + E = B + F = C + G = D + H = 56,$$
$$a + b + c + d = e + f + g + h = 18, \quad \text{and}$$
$$a + e = b + f = c + g = d + h = 9,$$

then the resultant square will meet all the above requirements except that of symmetry. It is not possible to assign values to the letters to make the square symmetrical without sacrificing other more important properties. It would also be possible to make the half-rows and half-columns magic by making

$$A + C + F + H = B + D + E + G = 112, \quad \text{and}$$
$$a + c + f + h = b + d + e + g = 18,$$

but if this condition is met it will not be possible to make the bent diagonals correct. It is thus confirmed that, while certain conditions can be met, it is not always possible to meet all of the possible combinations of these conditions.

The following is one of the many sets of numerical values which meet all of our original conditions except that of symmetry: $A = 8(E = 48)$, $B = 32(F = 24)$, $C = 16(G = 40)$, $D = 56(H = 0)$, $a = 1(e = 8)$, $b = 6(f = 3)$, $c = 4(g = 5)$, and $d = 7(h = 2)$. Substituting these values in Figure 48 gives Figure 49a.

56+8	0+6	56+5	0+7	56+1	0+3	56+4	0+2
40+1	16+3	40+4	16+2	40+8	16+6	40+5	16+7
32+8	24+6	32+5	24+7	32+1	24+3	32+4	24+2
48+1	8+3	48+4	8+2	48+8	8+6	48+5	8+7
0+8	56+6	0+5	56+7	0+1	56+3	0+4	56+2
16+1	40+3	16+4	40+2	16+8	40+6	16+5	40+7
24+8	32+6	24+5	32+7	24+1	32+3	24+4	32+2
8+1	48+3	8+4	48+2	8+8	48+6	8+5	48+7

Fig. 49a.

This square will, in turn, form Figure 49b when the numbers in each cell are added together. Examination will show that these squares do, in fact, have the remarkable properties claimed for them:

(a) The main square is pandiagonal.

(b) The bent diagonals are magic, whether drawn from the top, bottom, right-hand or left-hand side of the square.

64	6	61	7	57	3	60	2
41	19	44	18	48	22	45	23
40	30	37	31	33	27	36	26
49	11	52	10	56	14	53	15
8	62	5	63	1	59	4	58
17	43	20	42	24	46	21	47
32	38	29	39	25	35	28	34
9	51	12	50	16	54	13	55

Fig. 49b.   An eighth-order pandiagonal magic square with many additional magical properties.

(c) In each of the corner 4-by-4 subsquares the main diagonals and the broken diagonals are magic.

(d) Any 2-by-2 subsquare is magic.

(e) In fact, any four numbers symmetrically located with respect to the intersection of any line between two columns and any line between two rows are magic.

(f) Every one of the 49 possible $\{p,q\}$ series except $\{\pm 1,4\}$, $\{\pm 3,4\}$, $\{4,\pm 1\}$, $\{4,\pm 3\}$ is magic.

The above values were selected with a particular purpose in view. If we transpose rows 2 and 4, and 3 and 5, along with columns 2 and 3, and 3 and 5, we will convert Figure 49b into Figure 14 with its different, but also remarkable, properties. It follows that by a different rearrangement of the rows and columns we could generate Figure 11. In other words, we now see that the method described in Chapter 2 for constructing doubly-even-order magic squares also becomes a special case under our new cyclical method.

In closing this chapter we would like to call attention to one point (which we are sure the reader has already observed), namely, that while the conditions listed in this chapter will—when met—ensure that the resultant square will have the desired properties, they are by no means essential. Squares with the desired properties can be constructed which do not meet one, or more, of the requirements listed. On the other hand, the method does furnish a powerful tool for the construction of magic squares with added magical properties that can be predicted in advance.

11

_____

DOUBLY-EVEN-ORDER
MAGIC SQUARES

If you merely wish to construct a symmetrical magic square of the doubly-even order the method described in Part I is by far the simplest one. However, if you are interested in additional properties (particularly those of bimagic and trimagic squares as described in the next two chapters) the following method* is very important.

Before actually describing the process it will be necessary to explain the special nomenclature to be used. Since it simplifies the equations to be used, advantage will be taken of the fact that if each number in a magic square (or for that matter, in a bimagic or trimagic square) is increased by 1 the resultant square will also be magic (or bimagic or trimagic as the case may be). Thus, in all cases we will be generating squares formed of the integers $0, 1, 2, 3, \ldots, (n^2 - 1)$, inclusive, and then converting them to ordinary normal squares by adding 1 to each number.

In the last two chapters we have, without calling attention to the

_____

* This method and its modifications, as discussed in the following chapters, is an original one developed by the authors.

fact, been dealing with the numbers 1 to n^2 expressed to the base n. Here we will operate similarly on the numbers $0, 1, 2, 3, \ldots, (n^2 - 1)$, which—when expressed to the base n—become $0, 1, 2, \ldots, (n - 1)$; $n + 0, n + 1, n + 2, \ldots, n + (n - 1)$; $2n + 0, 2n + 1, 2n + 2, \ldots, 2n + (n - 1); \cdots; (n - 1)n + 0, (n - 1)n + 1, (n - 1)n + 2, \ldots, (n - 1)n + (n - 1)$. Any given number can thus be written to the base n as $vn + w$ where v and w go from 0 to $(n - 1)$, inclusive. Further let us define $-v$ as being shorthand for $(n - 1) - v$ and $-w$ as being shorthand for $(n - 1) - w$. As a further simplification let us agree to write $(vn + w)$ as v, w.

Using this nomenclature we have, when $n = 8$:

$$0, 1 = (0) \times 8 + 1 = 1$$

$$0, -1 = (0) \times 8 + (8 - 1) - 1 = 6$$

$$0, 0 = (0) \times 8 + 0 = 0$$

$$-0, -0 = [(8 - 1) - 0] \times 8 + (8 - 1) - 0 = 63$$

$$-2, 3 = [(8 - 1) - 2] \times 8 + 3 = 43$$

$$2, -3 = (2) \times 8 + (8 - 1) - 3 = 20$$

Notice also that in each case

$$v, w + -v, -w = v, -w + -v, w = (n^2 - 1).$$

To return to the actual construction of the square: consider the main n-by-n square as being made up of four $n/2$-by-$n/2$ subsquares where each subsquare is derived from a $n/2$-by-$n/2$ primary square in accordance with instructions contained in a 2-by-2 generating square. The method can be demonstrated by using the 2-by-2 generating square shown in Figure 50 and the 4-by-4 primary square shown in Figure 51a to generate an 8-by-8 pandiagonal magic square.

$+0, +0$	$-0, +0$
$+0, -0$	$-0, -0$

Fig. 50.   Generating square.

+3, +3	+2, −3	−1, +3	−0, −3
−3, +2	−2, −2	+1, +2	+0, −2
+3, −1	+2, +1	−1, −1	−0, +1
−3, −0	−2, +0	+1, −0	+0, +0

Fig. 51a.   Primary square.

The main 8-by-8 square is formed by placing the primary square in each of the four corners after modifying it in accordance with the instructions contained in the generating square. Examination of Figure 51b will make this clear. The subsquare in the upper left-hand corner is exactly like the primary square since the number in the upper left-hand corner of the generating square, +0, +0, directs that no change be made. The subsquare in the upper right-hand corner differs from the primary square in that the signs of all the "tens" numbers are reversed in accordance with the instructions in the upper right-hand corner of the generating square, −0, +0, which says to leave the signs of the "units" alone and reverse the signs of the "tens." In a similar manner the number in the lower left-hand corner of the generating square, +0, −0, says to leave the signs of the "tens" alone and to reverse the signs of the "units." Finally, the number in the lower right-hand corner of the generating square, −0, −0, says to reverse all signs in the primary square.

+3, +3	+2, −3	−1, +3	−0, −3	−3, +3	−2, −3	+1, +3	+0, −3
−3, +2	−2, −2	+1, +2	+0, −2	+3, +2	+2, −2	−1, +2	−0, −2
+3, −1	+2, +1	−1, −1	−0, +1	−3, −1	−2, +1	+1, −1	+0, +1
−3, −0	−2, +0	+1, −0	+0, +0	+3, −0	+2, +0	−1, −0	−0, +0
+3, −3	+2, +3	−1, −3	−0, +3	−3, −3	−2, +3	+1, −3	+0, +3
−3, −2	−2, +2	+1, −2	+0, +2	+3, −2	+2, +2	−1, −2	−0, +2
+3, +1	+2, −1	−1, +1	−0, −1	−3, +1	−2, −1	+1, +1	+0, −1
−3, +0	−2, −0	+1, +0	+0, −0	+3, +0	+2, −0	−1, +0	−0, −0

Fig. 51b.

The ease with which this square can be checked is readily apparent. The sum of the "tens" and "units" in each row, column, upward and downward diagonal, is zero! Figure 51c shows the same square converted to base 8 by adding $(n-1)=8-1=7$ to each negative number. While this square cannot be checked quite as simply as Figure 51b, it is still an easy task. The sum of the "tens" and "units" in each row, column, upward and downward diagonal, will total 28, as required.

3,3	2,4	6,3	7,4	4,3	5,4	1,3	0,4
4,2	5,5	1,2	0,5	3,2	2,5	6,2	7,5
3,6	2,1	6,6	7,1	4,6	5,1	1,6	0,1
4,7	5,0	1,7	0,0	3,7	2,0	6,7	7,0
3,4	2,3	6,4	7,3	4,4	5,3	1,4	0,3
4,5	5,2	1,5	0,2	3,5	2,2	6,5	7,2
3,1	2,6	6,1	7,6	4,1	5,6	1,1	0,6
4,0	5,7	1,0	0,7	3,0	2,7	6,0	7,7

Fig. 51c.

Figure 51d shows the same square converted to the normal base 10 by use of the relationship:

$$v,w = vn + w = (v) \times 8 + w.$$

27	20	51	60	35	44	11	4
34	45	10	5	26	21	50	61
30	17	54	57	38	41	14	1
39	40	15	0	31	16	55	56
28	19	52	59	36	43	12	3
37	42	13	2	29	18	53	58
25	22	49	62	33	46	9	6
32	47	8	7	24	23	48	63

Fig. 51d.

Figure 51e shows the final square constructed by adding 1 to each number in Figure 51d.

28	21	52	61	36	45	12	5
35	46	11	6	27	22	51	62
31	18	55	58	39	42	15	2
40	41	16	1	32	17	56	57
29	20	53	60	37	44	13	4
38	43	14	3	30	19	54	59
26	23	50	63	34	47	10	7
33	48	9	8	25	24	49	64

Fig. 51e.   An eighth-order pandiagonal magic square.

The above method of generating pandiagonal doubly-even-order magic squares, while demonstrated by a particular example, is perfectly general and may be described formally as follows:

CONSTRUCTION OF DOUBLY-EVEN-ORDER PANDIAGONAL MAGIC SQUARES

The method is based upon the generating square shown in Figure 50 and a primary square (with their numbers expressed in the "modified" nomenclature previously described) of the $n/2$-by-$n/2$ order, which meets Requirements A to E, inclusive:

REQUIREMENT A: Each of the $n^2/4$ possible combinations of $\pm 0, \pm 1, \pm 2, \ldots, \pm (n/2-1)$ in the "tens" place with $\pm 0, \pm 1, \pm 2, \ldots, \pm (n/2-1)$ in the "units" place must appear once, and only once. The actual sign of the numeral is unimportant, so long as the remaining conditions are met. The important point to watch is that there is no duplication. For example, if the combination $+3, -1$ appears in any cell in the primary square the combinations $+3, +1$, $-3, +1$, and $-3, -1$ must not appear anywhere else in the primary square.

REQUIREMENT B: Each row must contain $n/4$ positive and $n/4$ negative "units."

REQUIREMENT C: Each column must contain $n/4$ positive and $n/4$ negative "tens."

REQUIREMENT D: The sum of the "units" in each row must equal zero.

REQUIREMENT E: The sum of the "tens" in each column must equal zero.

Given the generating square, Figure 50, and a primary square which meets the above requirements, a pandiagonal doubly-even square of the nth order may be generated by following the steps listed below:

STEP 1: Place the primary square in each quarter of the main square following the instructions given by the generating square. That is, leave the signs as they are when the sign in the generating square is plus and reverse them when it is minus.

STEP 2: Convert the "modified" nomenclature into regular base n nomenclature by adding $(n-1)$ to each negative number.

STEP 3: Convert the base n nomenclature into regular base 10 nomenclature by adding the "units" numeral to n times the "tens" numeral.

STEP 4: Add 1 to each number in the square.

As another example, the 2-by-2 primary square shown in Figure 52a may be used to generate the fourth-order pandiagonal magic square shown in Figure 52b. Checking this square is quite simple. All the correct totals are zero. The totals of the final square, Figure 52c, will have to be magic (assuming no mistakes in arithmetic in transforming Figure 52b into the normal form).

$-1,-1$	$+0,+1$
$+1,+0$	$-0,-0$

Fig. 52a.

$-1,-1$	$+0,+1$	$+1,-1$	$-0,+1$
$+1,+0$	$-0,-0$	$-1,+0$	$+0,-0$
$-1,+1$	$+0,-1$	$+1,+1$	$-0,-1$
$+1,-0$	$-0,+0$	$-1,-0$	$+0,+0$

Fig. 52b.

11	2	7	14
5	16	9	4
10	3	6	15
8	13	12	1

Fig. 52c. A fourth-order pandiagonal magic square.

As the reader continues his investigation of doubly-even-order magic squares he will find that, just as the doubly-even-order square is easier to construct than the singly-even-order square, so doubly-even-order squares of the order 2^z, where z is greater than 2 (that is, $n = 8$, 16, 32 and so forth) can be constructed with many more properties than other doubly-even-order squares. The primary squares in Figures 51a and 52a were selected so that the final squares (Figures 51e and 52c) have the property that all 2-by-2 subsquares are magic. In addition, Figure 51e has the property that all bent diagonals are magic.

Just as the properties of Figure 11 were increased by rearranging the order of the rows and columns, Figure 51e can be improved by transposing rows 2 and 4 and rows 3 and 5, along with columns 2 and 4 and columns 3 and 5 to convert Figure 51e into Figure 51f, which has all the properties of Figure 14!

28	21	36	45	52	61	12	5
35	46	27	22	11	6	51	62
29	20	37	44	53	60	13	4
38	43	30	19	14	3	54	59
31	18	39	42	55	58	15	2
40	41	32	17	16	1	56	57
26	23	34	47	50	63	10	7
33	48	25	24	9	8	49	64

Fig. 51f. Another remarkable pandiagonal magic square
with all the magical properties of Fig. 14.

As a matter of possible interest Figure 51g shows Figure 51b after making the same interchange of rows and columns.

+3, +3	+2, -3	-3, +3	-2, -3	-1, +3	-0, -3	+1, +3	+0, -3
-3, +2	-2, -2	+3, +2	+2, -2	+1, +2	+0, -2	-1, +2	-0, -2
+3, -3	+2, +3	-3, -3	-2, +3	-1, -3	-0, +3	+1, -3	+0, +3
-3, -2	-2, +2	+3, -2	+2, +2	+1, -2	+0, +2	-1, -2	-0, +2
+3, -1	+2, +1	-3, -1	-2, +1	-1, -1	-0, +1	+1, -1	+0, +1
-3, -0	-2, +0	+3, -0	+2, +0	+1, -0	+0, +0	-1, -0	-0, +0
+3, +1	+2, -1	-3, +1	-2, -1	-1, +1	-0, -1	+1, +1	+0, -1
-3, +0	-2, -0	+3, +0	+2, -0	+1, +0	+0, -0	-1, +0	-0, -0

Fig. 51g.

Note how easy it is to check that this square does have all of the properties claimed for it when it is in the "modified" form. All correct totals are zero!

Actually this method is much more general than the above requirements would indicate. In fact, by using a more powerful generating square and the instructions employed in generating trimagic squares (as described later—actually the changes are very slight), it is possible to generate Figure 51g (and from it Figure 51f, of course) directly from the 4-by-4 generating square, Figure 53a, and the 2-by-2 primary square, Figure 53b.

+1, +1	-1, +1	-0, +1	+0, +1
+1, -1	-1, -1	-0, -1	+0, -1
+1, -0	-1, -0	-0, -0	+0, -0
+1, +0	-1, +0	-0, +0	+0, +0

Fig. 53a.   Generating square for Fig. 51g.

+1, +1	+0, -1
-1, +0	-0, -0

Fig. 53b.   Primary square for Fig. 51g.

12

BIMAGIC SQUARES

A magic square is said to be *bimagic* if, in addition to meeting the normal requirements for a magic square, the square formed by replacing each one of the n^2 numbers in the original square by its square is also magic. Thus the original square will consist of the numbers $1, 2, 3, \ldots, n^2$ and the second-degree square will consist of the numbers $1^2, 2^2, 3^2, \ldots, n^4$. The magic constant of the first-degree square is, of course, $n(n^2+1)/2$ and of the second-degree square, $n(n^2+1)(2n^2+1)/6$.

The construction of bimagic squares is merely an extension of the method employed in the previous chapter to generate pandiagonal doubly-even-order magic squares, the only difference being that the primary square must meet the following additional requirements:

REQUIREMENT F: In each row and in each column the sign of the "tens" number and the "units" number must be the same in one-half of the cells and different in the remaining half.

REQUIREMENT G: For both main diagonals one-fourth of the cells must have both the signs of the "tens" number and the "units" number positive, one-fourth must have both signs negative, one-fourth must have the "tens" number positive and the "units" number negative, and the remaining fourth of the cells must have the "tens" number negative and the "units" number positive.

REQUIREMENT H: The numbers in the "tens" and "units" place in each row, column and main diagonal (hereafter referred to as each rcd) must consist of $0, 1, 2, \ldots, (n/2 - 1)$ in some order. The sign of the number is immaterial as long as the other requirements are met.

REQUIREMENT I: For both main diagonals the sum of the "tens" numbers opposite negative "units" numbers, minus the sum of the "tens" numbers opposite positive "units" numbers, plus the sum of the "units" numbers opposite negative "tens" numbers, minus the sum of the "units" numbers opposite positive "tens" numbers, must equal zero.

REQUIREMENT J: For both main diagonals the sum of the cross-products of the "tens" number and the "units" number in the same cell must be zero.

Figure 54a is such a square. That it meets requirements A to H, inclusive, is readily apparent. In the case of requirement I:

$$(3-2)-(0-1)+(2-1)-(3-0)=0,$$

and

$$(3-2)-(0-1)+(0-3)-(1-2)=0,$$

as required. Also, in the case of Requirement J:

$$(-2)(-1)+(-1)(+2)+(+3)(-0)+(+0)(+3)=0,$$

and

$$(+3)(-2)+(+0)(+1)+(-2)(-3)+(-1)(+0)=0,$$

as required.

$-2,-1$	$+3,+3$	$+0,-2$	$-1,+0$
$+0,-0$	$-1,+2$	$-2,-3$	$+3,+1$
$-1,-3$	$+0,+1$	$+3,-0$	$-2,+2$
$+3,-2$	$-2,+0$	$-1,-1$	$+0,+3$

Fig. 54a.   Primary square for Fig. 54b.

Using this primary square and Figure 50 as a generating square, it is a simple matter to construct Figure 54b in the "modified" nomenclature. It is in this form that the ordinary properties of the first-degree square and the cross-products are checked most easily. There are an equal number of positive and negative terms in each rcd. The sum of the "tens" numbers, the "units" numbers and the cross-products of each rcd is zero as required for the main square to be correct. Also, the sum of the "tens" numbers and the "units" numbers of all upward and downward broken diagonals is zero. The square is, therefore, not only magic but also pandiagonal in the first degree.

$-2,-1$	$+3,+3$	$+0,-2$	$-1,+0$	$+2,-1$	$-3,+3$	$-0,-2$	$+1,+0$
$+0,-0$	$-1,+2$	$-2,-3$	$+3,+1$	$-0,-0$	$+1,+2$	$+2,-3$	$-3,+1$
$-1,-3$	$+0,+1$	$+3,-0$	$-2,+2$	$+1,-3$	$-0,+1$	$-3,-0$	$+2,+2$
$+3,-2$	$-2,+0$	$-1,-1$	$+0,+3$	$-3,-2$	$+2,+0$	$+1,-1$	$-0,+3$
$-2,+1$	$+3,-3$	$+0,+2$	$-1,-0$	$+2,+1$	$-3,-3$	$-0,+2$	$+1,-0$
$+0,+0$	$-1,-2$	$-2,+3$	$+3,-1$	$-0,+0$	$+1,-2$	$+2,+3$	$-3,-1$
$-1,+3$	$+0,-1$	$+3,+0$	$-2,-2$	$+1,+3$	$-0,-1$	$-3,+0$	$+2,-2$
$+3,+2$	$-2,-0$	$-1,+1$	$+0,-3$	$-3,+2$	$+2,-0$	$+1,+1$	$-0,-3$

Fig. 54b.

Figure 54c shows the same square converted to base 8 by adding $(n-1)=(8-1)=7$ to each negative number. Figure 54c is the most convenient form for checking the bimagic properties of the "tens" and "units" numbers. Note that the "tens" and "units" numbers of each rcd are composed of one of the following combinations of numbers:

0, 1, 2, 3, 4, 5, 6 and 7,

0, 3, 5, 6, 0, 3, 5 and 6, or

1, 2, 4, 7, 1, 2, 4 and 7.

Since:

$$0+1+4+9+16+25+36+49=140,$$

$$0+9+25+36+0+9+25+36=140, \text{ and}$$

$$1+4+16+49+1+4+16+49=140,$$

as required, it is clear that the sum of the squares of the "tens" and "units" will be correct in each case.

5,6	3,3	0,5	6,0	2,6	4,3	7,5	1,0
0,7	6,2	5,4	3,1	7,7	1,2	2,4	4,1
6,4	0,1	3,7	5,2	1,4	7,1	4,7	2,2
3,5	5,0	6,6	0,3	4,5	2,0	1,6	7,3
5,1	3,4	0,2	6,7	2,1	4,4	7,2	1,7
0,0	6,5	5,3	3,6	7,0	1,5	2,3	4,6
6,3	0,6	3,0	5,5	1,3	7,6	4,0	2,5
3,2	5,7	6,1	0,4	4,2	2,7	1,1	7,4

Fig. 54c.

If you desire to use this square to check the cross-products, the correct total is 98. For example, take row 2: $(0)(0)+(6)(5)+(5)(3)+(3)(6)+(7)(0)+(1)(5)+(2)(3)+(4)(6)=98$, as required.

There remains only the task of converting this square to the base 10 and adding 1 to each number to get Figure 54d, which is pandiagonal in the first degree and magic in the second degree.

In other words, Figure 54d is an eighth-order bimagic square which is also pandiagonal in the first degree.

47	28	6	49	23	36	62	9
8	51	45	26	64	11	21	34
53	2	32	43	13	58	40	19
30	41	55	4	38	17	15	60
42	29	3	56	18	37	59	16
1	54	44	31	57	14	20	39
52	7	25	46	12	63	33	22
27	48	50	5	35	24	10	61

Fig. 54d.   A pandiagonal bimagic square of the eighth order.

To show that the above figures are correct is a simple matter. Note that:

$$140 \times 8^2 + 2 \times 8 \times 98 + 140 = 10{,}668,$$

which is the magic constant for a square formed of the numbers $0^2, 1^2, 2^2, 3^2, \ldots, (8^2-1)^2$, inclusive.

It is believed that bimagic squares of an order less than eight do not exist. Figure 55a is a symmetrical bimagic square of the ninth order expressed to the base 9 for ease in checking its properties. The correct sum of the cross-products is 144, and for the sum of the squares of the "tens" and of the "units" it is 204. This gives us:

$$204^2 \times 9^2 + 2 \times 9 \times 144 + 204 = 19{,}320,$$

the magic constant for a ninth-order, second-degree square formed of the numbers $0^2, 1^2, 2^2, 3^2, \ldots, (9^2-1)^2$, inclusive.

8,7	7,3	6,2	2,0	1,8	0,4	5,5	4,1	3,6
3,4	5,0	4,8	6,6	8,5	7,1	0,2	2,7	1,3
1,1	0,6	2,5	4,3	3,2	5,7	7,8	6,4	8,0
4,6	3,5	5,1	7,2	6,7	8,3	1,4	0,0	2,8
2,3	1,2	0,7	5,8	4,4	3,0	8,1	7,6	6,5
6,0	8,8	7,4	0,5	2,1	1,6	3,7	5,3	4,2
0,8	2,4	1,0	3,1	5,6	4,5	6,3	8,2	7,7
7,5	6,1	8,6	1,7	0,3	2,2	4,0	3,8	5,4
5,2	4,7	3,3	8,4	7,0	6,8	2,6	1,5	0,1

Fig. 55a.   A ninth-order, symmetrical, bimagic square.

Needless to say, Figure 55a was not constructed by the method described earlier in this chapter. It is included here as a matter of possible interest.

Actually Figure 55a was generated from Figure 55b by letting $A = 6$, $B = 0$, $C = 3$, $a = 2$, $b = 1$, $c = 0$, $\mathbf{A} = 0$, $\mathbf{B} = 3$, $\mathbf{C} = 6$, $\mathbf{a} = 0$, $\mathbf{b} = 1$ and $\mathbf{c} = 2$. In fact, Figure 55b will be a bimagic square of the ninth order whenever A, B, C equals 0, 3, 6 in some order and a, b, c equals 0, 1, 2 in some order (or vice versa) and $\mathbf{A}$, $\mathbf{B}$, $\mathbf{C}$ equals 0, 3,

6 in some order and **a**, **b**, **c** equals 0, 1, 2 in some order (or vice versa). If care is taken to assign values properly, the square can be made (as in the case of Figure 55a) symmetrical.

A+a	*A+b*	*A+c*	*B+a*	*B+b*	*B+c*	*C+a*	*C+b*	*C+c*
C+b	**B+a**	**A+c**	**A+a**	**C+c**	**B+b**	**B+c**	**A+b**	**C+a**
C+c	*C+a*	*C+b*	*A+c*	*A+a*	*A+b*	*B+c*	*B+a*	*B+b*
B+b	**A+a**	**C+c**	**C+a**	**B+c**	**A+b**	**A+c**	**C+b**	**B+a**
B+b	*B+c*	*B+a*	*C+b*	*C+c*	*C+a*	*A+b*	*A+c*	*A+a*
A+b	**C+a**	**B+c**	**B+a**	**A+c**	**C+b**	**C+c**	**B+b**	**A+a**
C+b	*C+c*	*C+a*	*A+b*	*A+c*	*A+a*	*B+b*	*B+c*	*B+a*
C+a	**B+c**	**A+b**	**A+c**	**C+b**	**B+a**	**B+b**	**A+a**	**C+c**
B+a	*B+b*	*B+c*	*C+a*	*C+b*	*C+c*	*A+a*	*A+b*	*A+c*
B+a	**A+c**	**C+b**	**C+c**	**B+b**	**A+a**	**A+b**	**C+a**	**B+c**
A+c	*A+a*	*A+b*	*B+c*	*B+a*	*B+b*	*C+c*	*C+a*	*C+b*
A+a	**C+c**	**B+b**	**B+c**	**A+b**	**C+a**	**C+b**	**B+a**	**A+c**
B+c	*B+a*	*B+b*	*C+c*	*C+a*	*C+b*	*A+c*	*A+a*	*A+b*
C+c	**B+b**	**A+a**	**A+b**	**C+a**	**B+c**	**B+a**	**A+c**	**C+b**
A+b	*A+c*	*A+a*	*B+b*	*B+c*	*B+a*	*C+b*	*C+c*	*C+a*
B+c	**A+b**	**C+a**	**C+b**	**B+a**	**A+c**	**A+a**	**C+c**	**B+b**
C+a	*C+b*	*C+c*	*A+a*	*A+b*	*A+c*	*B+a*	*B+b*	*B+c*
A+c	**C+b**	**B+a**	**B+b**	**A+a**	**C+c**	**C+a**	**B+c**	**A+b**

Fig. 55b.

13

TRIMAGIC SQUARES

A magic square is said to be *trimagic* when, in addition to meeting the requirements for a bimagic square, the square formed by replacing every number by its cube is also magic. The magic constant of a third-degree square is $(n^3/4)(n^2+1)^2$.

The construction of a trimagic square is much more involved than any square we have considered so far. W. W. R. Ball* points out that the "construction of squares which shall be magic for the original numbers, for their squares, and for their cubes has also been studied. I know of no square of this kind which is of a lower order than 64, and the construction of a square of that order is not a 'recreation.'"

Maurice Kraitchik† informs us that "Tarry was the first to give a method of forming a trimagic square of order 128. His method was improved by Cazalas, who formed trimagic squares of orders 64 and

---

* W. W. R. Ball, *op. cit.*, p. 213.
† *Mathematical Recreations* by Maurice Kraitchik, Dover Publications, Inc., pp. 176-7.

81. Mr. Royal V. Heath has formed many bimagic squares, and a trimagic square of order 64 different from Cazalas' square."

The authors, using the generating square given in Figure 56 and the primary square given in Figure 57, constructed the trimagic square of the 32nd order shown in Figure 58. So far as is known, this is the first trimagic square ever to be constructed of an order lower than 64. It has been completely checked by the use of IBM equipment and proved to be correct. The method is perfectly general and flexible. Any number of trimagic squares of the 32nd (64th, 128th, etc.) order can be constructed by its use.

+0, +0	−1, +0	+0, −1	−1, −1
+1, −1	−0, −1	+1, +0	−0, +0
−1, +1	+0, +1	−1, −0	+0, −0
−0, −0	+1, −0	−0, +1	+1, +1

Fig. 56. The generating square used in the construction of the 32nd-order trimagic square shown in Fig. 58.

+0, +1	−6, −1	−0, −7	+6, +7	−3, +2	+5, −2	+3, −4	−5, +4
+1, −5	−7, +5	−1, +3	+7, −3	−2, −6	+4, +6	+2, +0	−4, −0
−2, +2	+4, −2	+2, −4	−4, +4	+1, +1	−7, −1	−1, −7	+7, +7
−3, −6	+5, +6	+3, +0	−5, −0	+0, −5	−6, +5	−0, +3	+6, −3
−7, −0	+1, +0	+7, +6	−1, −6	+4, −3	−2, +3	−4, +5	+2, −5
−6, +4	+0, −4	+6, −2	−0, +2	+5, +7	−3, −7	−5, −1	+3, +1
+5, −3	−3, +3	−5, +5	+3, −5	−6, −0	+0, +0	+6, +6	−0, −6
+4, +7	−2, −7	−4, −1	+2, +1	−7, +4	+1, −4	+7, −2	−1, +2

Fig. 57. The primary square used in the construction of the 32nd-order trimagic square shown in Fig. 58.

It is believed worthy of note that by this method it is possible for a 32nd-order trimagic square to be uniquely expressed in terms of one 8-by-8 primary square and one 4-by-4 generating square, neither of which contains any number greater than 7.

What is more, it is possible to prove that the generated 32nd-order square will be trimagic without actually constructing the square and testing it!

2	831	1017	200	899	190	124	837	738	479	281	552	355	606	668	421
59	774	964	253	954	135	65	896	731	486	292	541	346	615	673	416
931	158	92	869	34	799	985	232	323	638	700	389	706	511	313	520
922	167	97	864	27	806	996	221	378	583	641	448	763	454	260	573
800	33	231	986	157	932	870	91	512	705	519	314	637	324	390	699
805	28	222	995	168	921	863	98	453	764	574	259	584	377	447	642
189	900	838	123	832	1	199	1018	605	356	422	667	480	737	551	282
136	953	895	66	773	60	254	963	616	345	415	674	485	732	542	291
279	554	752	465	662	427	365	596	1015	202	16	817	118	843	909	180
302	531	725	492	687	402	344	617	974	243	53	780	79	882	952	137
694	395	333	628	311	522	720	497	86	875	941	148	983	234	48	785
655	434	376	585	270	563	757	460	111	850	920	169	1006	211	21	812
521	312	498	719	396	693	627	334	233	984	786	47	876	85	147	942
564	269	459	758	433	656	586	375	212	1005	811	22	849	112	170	919
428	661	595	366	553	280	466	751	844	117	179	910	201	1016	818	15
401	688	618	343	532	301	491	726	881	80	138	951	244	973	779	54

746	471	273	560	363	598	660	429	10	823	1009	208	907	182	116	845
723	494	300	533	338	623	681	408	51	782	972	245	946	143	73	888
331	630	692	397	714	503	305	528	939	150	84	877	42	791	977	240
370	591	649	440	755	462	268	565	914	175	105	856	19	814	1004	213
504	713	527	306	629	332	398	691	792	41	239	978	149	940	878	83
461	756	566	267	592	369	439	650	813	20	214	1003	176	913	855	106
597	364	430	659	472	745	559	274	181	908	846	115	824	9	207	1010
624	337	407	682	493	724	534	299	144	945	887	74	781	52	246	971
1023	194	8	825	126	835	901	188	287	546	744	473	670	419	357	604
966	251	61	772	71	890	960	129	294	539	733	484	679	410	352	609
94	867	933	156	991	226	40	793	702	387	325	636	319	514	712	505
103	858	928	161	998	219	29	804	647	442	384	577	262	571	765	452
225	992	794	39	868	93	155	934	513	320	506	711	388	701	635	326
220	997	803	30	857	104	162	927	572	261	451	766	441	648	578	383
836	125	187	902	193	1024	826	7	420	669	603	358	545	288	474	743
889	72	130	959	252	965	771	62	409	680	610	351	540	293	483	734

Fig. 58. A 32nd-order trimagic square.

23	810	1008	209	918	171	109	852	759	458	272	561	374	587	653	436
46	787	981	236	943	146	88	873	718	499	309	524	335	626	696	393
950	139	77	884	55	778	976	241	342	619	685	404	727	490	304	529
911	178	120	841	14	819	1013	204	367	594	664	425	750	467	277	556
777	56	242	975	140	949	883	78	489	728	530	303	620	341	403	686
820	13	203	1014	177	912	842	119	468	749	555	278	593	368	426	663
172	917	851	110	809	24	210	1007	588	373	435	654	457	760	562	271
145	944	874	87	788	45	235	982	625	336	394	695	500	717	523	310
258	575	761	456	643	446	380	581	994	223	25	808	99	862	924	165
315	518	708	509	698	391	321	640	987	230	36	797	90	871	929	160
675	414	348	613	290	543	729	488	67	894	956	133	962	255	57	776
666	423	353	608	283	550	740	477	122	839	897	192	1019	198	4	829
544	289	487	730	413	676	614	347	256	961	775	58	893	68	134	955
549	284	478	739	424	665	607	354	197	1020	830	3	840	121	191	898
445	644	582	379	576	257	455	762	861	100	166	923	224	993	807	26
392	697	639	322	517	316	510	707	872	89	159	930	229	988	798	35

767	450	264	569	382	579	645	444	31	802	1000	217	926	163	101	860
710	507	317	516	327	634	704	385	38	795	989	228	935	154	96	865
350	611	677	412	735	482	296	537	958	131	69	892	63	770	968	249
359	602	672	417	742	475	285	548	903	186	128	833	6	827	1021	196
481	736	538	295	612	349	411	678	769	64	250	967	132	957	891	70
476	741	547	286	601	360	418	671	828	5	195	1022	185	904	834	127
580	381	443	646	449	768	570	263	164	925	859	102	801	32	218	999
633	328	386	703	508	709	515	318	153	936	866	95	796	37	227	990
1002	215	17	816	107	854	916	173	266	567	753	464	651	438	372	589
979	238	44	789	82	879	937	152	307	526	716	501	690	399	329	632
75	886	948	141	970	247	49	784	683	406	340	621	298	535	721	496
114	847	905	184	1011	206	12	821	658	431	361	600	275	558	748	469
248	969	783	50	885	76	142	947	536	297	495	722	405	684	622	339
205	1012	822	11	848	113	183	906	557	276	470	747	432	657	599	362
853	108	174	915	216	1001	815	18	437	652	590	371	568	265	463	754
880	81	151	938	237	980	790	43	400	689	631	330	525	308	502	715

The following data are furnished for the use of any reader who may desire to make his own check:

(1) The sum of the numbers in any row, column or main diagonal is equal to 16,400.

(2) The sum of the squares of the numbers in any row, column or main diagonal is equal to 11,201,200.

(3) The sum of the cubes of the numbers in any row, column or main diagonal is equal to 8,606,720,000.

The construction of trimagic squares involves the use of larger generating squares than we have used to date. This in turn requires an amplification of the instructions governing the actual formation of the main square.

The main square may be considered as being made up of subsquares, each subsquare being equivalent to the primary square after it has been modified in accordance with the instructions contained in the corresponding cell in the generating square. Obviously, there will be as many subsquares as there are cells in the generating square. In the case of Figures 56 and 57 the generating square is 4-by-4 and the primary square is 8-by-8. This means that there will be 4 times 4, or 16, subsquares in the main square and that the latter will be of the 4 times 8, or 32nd order.

Perhaps the easiest way to explain the process is to examine a few examples. The upper left-hand corner of the generating square is $+0, +0$. This says to place the primary square in the upper left-hand corner of the main square without changing anything. The next symbol in the generating square is $-1, +0$. This says to increase the numerical value of the number in the "tens" place by j (where j is the order of the primary square—8 in this case) times the "tens" number in the generating-square instruction (1 in this case) and then change the signs of all "tens" numbers. The $+0$ says to leave the "units" number alone. To be more specific, the top row of the primary square is:

$+0, +1$ $-6, -1$ $-0, -7$ $+6, +7$ $-3, +2$

$+5, -2$ $+3, -4$ $-5, +4$.

The instruction $-1, +0$ would change this to:

$-8, +1$ $+14, -1$ $+8, -7$ $-14, +7$ $+11, +2$

$-13, -2$ $-11, -4$ $+13, +4$.

Adding $(n-1) = (32-1) = 31$ to each negative number makes it:

23,1   14,30   8,24   17,7   11,2   18,29   20,27   13,4.

Converting these to the base 10 by adding 32 times the "tens" number to the "units" number will give us:

737   478   280   551   354   605   667   420

and adding 1 to each number will give:

738   479   281   552   355   606   668   421.

Examination of Figure 58 will show that these are, in fact, the numbers appearing in the top row in columns 8 to 15.

In other words, if we represent any number in the main square by $\pm v, \pm w$, in the primary square by $\pm x, \pm y$, and in the generating square by $\pm a, \pm b$, the numerical value of v and w is:

$$v = ja + x \quad \text{and} \quad w = jb + y.$$

Remember that we pay no attention to the signs of the various numbers involved in determining the numerical values of v and w. The numerical values having been determined, the sign prefixed to v is the same as that of x when a is positive and the opposite to that of x when a is negative. Likewise, the sign of w agrees with that of y when b is positive and is the opposite to that of y when b is negative.

So much for the actual construction of the final square from any given pair of generating and primary squares. The next step is to investigate the requirements to be met by these two squares if the final square is to have the desired properties. The requirements to be met to ensure that the main square will contain the numbers 0 to (n^2-1), inclusive, each one once and only once, are quite simple. They are listed in Table 1.

TABLE 1

PRIMARY SQUARE: The j-by-j primary square must contain each of the j^2 possible combinations of $x = 0, 1, 2, \ldots, x, \ldots, (j-1)$ (either positive or negative) and $y = 0, 1, 2, \ldots, y, \ldots, (j-1)$ (either positive or negative), once and only once.

GENERATING SQUARE: The k-by-k generating square must contain each of the k^2 possible combinations of $a = 0, 1, 2, \ldots, a, \ldots, (k/2-1), -(k/2-1), \ldots, -a, \ldots, -2, -1, -0$ and $b = 0, 1, 2, \ldots, b, \ldots, (k/2-1), -(k/2-1), \ldots, -b, \ldots, -2, -1, -0$, once and only once.

Examination of Figures 56 and 57 will show that they do meet these requirements.

Tables 2, 3 and 4 list the additional requirements which, if met, will ensure that the main square will be trimagic. Note particularly that where a choice of requirements is indicated, only one of the alternatives need be met. It is not necessary to meet them all. Note also that it is not necessary that the columns and/or main diagonals meet the same set of requirements that the rows meet.

As a matter of convenience certain symbols will be assigned special meanings:

Σ ( ) to mean the sum of the quantity in the ( ). Where an additional symbol is placed before the ( ) it qualifies which of the various possible quantities are to be summed, for example:

$\Sigma + y(x)$ would indicate the sum of all x opposite positive y.

TABLE 2

Requirement Number	Primary Square	Generating Square
2 A	$\Sigma(\lvert x\rvert)=\Sigma(\lvert y\rvert)$ $=j(j-1)/2$	
2 B		$\Sigma(\lvert a\rvert)=\Sigma(\lvert b\rvert)$ $=k(k/2-1)/2$
2 C	$\Sigma(x^2)=\Sigma(y^2)$ $=j(j-1)(2j-1)/6$	
2 D		$\Sigma(a^2)=\Sigma(b^2)$ $=k(k/2-1)(k-1)/6$
2 E	$\Sigma(x)=0$	or $\Sigma(a)=0$
2 F	$\Sigma(y)=0$	or $\Sigma(b)=0$
2 G	$\Sigma+y(\lvert x\rvert)=\Sigma-y(\lvert x\rvert)$	or $\Sigma+b(\lvert a\rvert)=\Sigma-b(\lvert a\rvert)$
2 H	$\Sigma+x(\lvert y\rvert)=\Sigma-x(\lvert y\rvert)$	or $\Sigma+a(\lvert b\rvert)=\Sigma-a(\lvert b\rvert)$
2 I(1)	$\Sigma+,+(\lvert x\rvert)-\Sigma+,-(\lvert x\rvert)$ $=\Sigma-,+(\lvert x\rvert)-\Sigma-,-(\lvert x\rvert)=0,$ or	
(2)		$\Sigma+a(b)=\Sigma-a(b)=0,$ or
(3)	$\Sigma+y(\lvert x\rvert)=\Sigma-y(\lvert x\rvert)$	and $\Sigma+a(b)=\Sigma-a(b),$ or
(4)	$\Sigma+y(x)=\Sigma-y(x)$	and $\Sigma(b)=0$
2 J(1)	$\Sigma+,+(\lvert y\rvert)-\Sigma-,+(\lvert y\rvert)$ $\Sigma+,-(\lvert y\rvert)-\Sigma-,-(\lvert y\rvert)=0,$ or	
(2)		$\Sigma+b(a)=\Sigma-b(a)=0,$ or
(3)	$\Sigma+x(\lvert y\rvert)=\Sigma-x(\lvert y\rvert)$	and $\Sigma+b(a)=\Sigma-b(a),$ or
(4)	$\Sigma+x(y)=\Sigma-x(y)$	and $\Sigma(a)=0$

TABLE 2–Continued

Requirement Number	Primary Square	Generating Square
2 K(1)	$\Sigma + x(y) = \Sigma - x(y) = 0$, or	
(2)		$\Sigma +, +(|a|) - \Sigma +, -(|a|)$ $= \Sigma -, +(|a|) - \Sigma -, -(|a|) = 0$, or
(3)	$\Sigma + x(y) = \Sigma - x(y)$	and $\Sigma + b(|a|) = \Sigma - b(|a|)$, or
(4)	$\Sigma(y) = 0$	and $\Sigma + b(a) = \Sigma - b(a)$
2 L(1)	$\Sigma + y(x) = \Sigma - y(x) = 0$, or	
(2)		$\Sigma +, +(|b|) - \Sigma -, +(|b|)$ $= \Sigma +, -(|b|) - \Sigma -, -(|b|) = 0$, or
(3)	$\Sigma + y(x) = \Sigma - y(x)$	and $\Sigma + a(|b|) = \Sigma - a(|b|)$, or
(4)	$\Sigma(x) = 0$	and $\Sigma + a(b) = \Sigma - a(b)$
2 M	$\Sigma + y(|xy|) = \Sigma - y(|xy|)$	or $\Sigma + b(|a|) = \Sigma - b(|a|)$
2 N	$\Sigma + x(|xy|) = \Sigma - x(|xy|)$	or $\Sigma + a(|b|) = \Sigma - a(|b|)$
2 O	$\Sigma + y(|x|) = \Sigma - y(|x|)$	or $\Sigma + b(|ab|) = \Sigma - b(|ab|)$
2 P	$\Sigma + x(|y|) = \Sigma - x(|y|)$	or $\Sigma + a(|ab|) = \Sigma - a(|ab|)$
2 Q	$\Sigma(y) = 0$	or $\Sigma + b(a^2) = \Sigma - b(a^2)$
2 R	$\Sigma(x) = 0$	or $\Sigma + a(b^2) = \Sigma - a(b^2)$
2 S	$\Sigma(x) = 0$	or $\Sigma + a(a^2) = \Sigma - a(a^2)$
2 T	$\Sigma(y) = 0$	or $\Sigma + b(b^2) = \Sigma - b(b^2)$
2 U	$\Sigma + y(x^2) = \Sigma - y(x^2)$	or $\Sigma(b) = 0$
2 V	$\Sigma + x(y^2) = \Sigma - x(y^2)$	or $\Sigma(a) = 0$
2 W	$\Sigma + x(x^2) = \Sigma - x(x^2)$	or $\Sigma(a) = 0$
2 X	$\Sigma + y(y^2) = \Sigma - y(y^2)$	or $\Sigma(b) = 0$

$\Sigma + xy(x)$ would indicate the sum of all x in cells where the product xy is positive.

$\Sigma +, -(x)$ would indicate the sum of all x in cells where the "tens" number is positive and the "units" number is negative.

$| \ |$ is used in its normal mathematical sense to mean the absolute value of the quantity between the parallel lines, that is, $|+h| = +h$ and $|-h| = +h$, thus

$\Sigma + a(|b|)$ is the sum of all b's opposite positive a's ignoring the signs of the b's.

Table 2 is applicable to the rows (columns or main diagonals) when each row (column or main diagonal), in both the primary and generating squares meets the following requirement—which we shall identify as Distribution A—one-fourth of the cells shall have both the "tens" and the "units" number positive, one-fourth shall have both signs negative, one-fourth shall have the "tens" number positive and the "units" number negative, and the remaining fourth shall have the "tens" number negative and the "units" number positive.

Examination of Figures 56 and 57 will show that this table is applicable to the rows and columns but not to the main diagonals.

Table 3 is applicable to the rows (columns or main diagonals) when each row (column or main diagonal) in the primary square has Distribution A and each row (column or main diagonal) in the generating square meets the following distribution requirement—which we shall identify as Distribution B—either one-half of the cells shall have both the "tens" number and the "units" number positive and the remaining half shall have both signs negative, or one-half of the cells shall have the "tens" number positive and the "units" number negative and the remaining half shall have the "tens" number negative and the "units" number positive.

Examination of Figures 56 and 57 will show that this table is applicable to the main diagonals.

TABLE 3

The requirements of Table 3 are the same as those of Table 2 plus the following additional requirements:

Requirement Number	Primary Square
3 A	$\Sigma + xy(|x|) = \Sigma - xy(|x|) = \Sigma + xy(|y|) = \Sigma - xy(|y|) = j(j-1)/4$
3 Y	$\Sigma + xy(|xy|) = \Sigma - xy(|xy|)$ or same thing $\Sigma(xy) = 0$

TABLE 4

The requirements of Table 4 are the same as those of Table 2 plus the following additional requirements:

Requirement Number	Generating Square
4 B	$\Sigma + ab(|a|) = \Sigma - ab(|a|) = \Sigma + ab(|b|) = \Sigma - ab(|b|) = k(k/2-1)/4$
4 Z	$\Sigma(ab) = 0$ or same thing $\Sigma + ab(|ab|) = \Sigma - ab(|ab|)$

Table 4 is applicable to the rows (columns or main diagonals) when each row (column or main diagonal) in the primary square has Distribution B and each row (column or main diagonal) in the generating square has Distribution A.

Examination of Figures 56 and 57 will show that this table does not apply to them.

14

IRREGULAR PANDIAGONAL
MAGIC SQUARES

Irregular pandiagonal magic squares can be defined as pandiagonal magic squares where the lower-case (or capital) letters in the intermediate square (from which the numerical square can be generated) do not appear an equal number of times in each row (column, or diagonal).

There are no irregular pandiagonal magic squares of any order less than seven. Professor Albert L. Candy* describes a method of construction of 640,120,320 magic squares of this type. This sounds like, and is, an astonishingly large number of seventh-order irregular pandiagonal magic squares (there are only 38,102,400 regular pandiagonal magic squares of this order). The really remarkable fact remains, however, that when the powerful tool of the new cyclical-method intermediate square is used these 640 million (plus) squares can all be constructed from only two basic squares, those shown in Figures 59 and 60.

---

* *Pandiagonal Magic Squares of Prime Order* by Albert L. Candy, published by the author, Nebraska, pp. 19 ff.

There is no doubt about these squares being irregular. Note that the letter *a* appears twice in the top row, and not at all in the next to top row, of both squares. Similar duplications appear elsewhere.

Examination of these two squares will show that they are both pandiagonal magic squares as far as the capital letters are concerned and that they will also be pandiagonal as far as the lower-case letters are concerned provided that:

$$(a-b)=(c-d)=(e-f).$$

$A+a$	$B+a$	$C+e$	$D+d$	$E+g$	$F+d$	$G+f$
$E+c$	$F+f$	$G+b$	$A+b$	$B+e$	$C+c$	$D+g$
$B+d$	$C+g$	$D+c$	$E+e$	$F+b$	$G+a$	$A+f$
$F+a$	$G+e$	$A+d$	$B+g$	$C+d$	$D+e$	$E+b$
$C+f$	$D+b$	$E+a$	$F+e$	$G+d$	$A+g$	$B+c$
$G+g$	$A+c$	$B+f$	$C+b$	$D+a$	$E+f$	$F+c$
$D+f$	$E+d$	$F+g$	$G+c$	$A+e$	$B+b$	$C+a$

Fig. 59.   Irregular pandiagonal magic square number 1—seventh order.

$A+a$	$B+a$	$C+f$	$D+d$	$E+g$	$F+d$	$G+e$
$E+d$	$F+e$	$G+b$	$A+b$	$B+e$	$C+c$	$D+g$
$B+d$	$C+g$	$D+c$	$E+e$	$F+b$	$G+a$	$A+f$
$F+a$	$G+f$	$A+c$	$B+g$	$C+d$	$D+e$	$E+b$
$C+e$	$D+b$	$E+a$	$F+f$	$G+d$	$A+g$	$B+c$
$G+g$	$A+d$	$B+f$	$C+a$	$D+a$	$E+f$	$F+c$
$D+f$	$E+c$	$F+g$	$G+c$	$A+e$	$B+b$	$C+b$

Fig. 60.   Irregular pandiagonal magic square number 2—seventh order.

Also note that each of the possible 49 combinations of the capital and the lower-case letters appears once and only once. It follows that the squares will be normal provided the capital letters equal 0,

7, 14, 21, 28, 35 and 42, in any order, and the lower-case letters equal 1, 2, 3, 4, 5, 6 and 7, in any order—as long as $(a-b)=(c-d)=(e-f)$, or vice versa.

Professor Candy also showed how to extend his method to include the construction of irregular pandiagonal magic squares of prime orders above seven. In each case it is possible, of course, to present them in the form of intermediate squares.

15

NONCYCLICAL PANDIAGONAL
MAGIC SQUARES

It is obvious that any irregular pandiagonal magic square is non-cyclical where, by definition, a noncyclical pandiagonal magic square is a pandiagonal magic square that does not have both the capital letters and lower-case letters in each row, column and diagonal cyclically arranged. The question remains, "Are there noncyclical regular pandiagonal magic squares when n is a prime number?"

As shown by Erich Stern,* the answer is, "Yes!" It is possible, if you meet the requirements listed below, to construct a normal cyclical pandiagonal magic square in which you can transpose two pairs of columns (rows or diagonals)—the second of which will depend upon the first pair selected—and still have the square remain pandiagonal. This transposition will, of course, destroy the

---

*"Number of Magic Squares Belonging to Certain Classes" by Erich Stern (translated from the German by W. R. Transue), *American Mathematical Monthly*, 1939, pp. 572 ff.

cyclical properties of the rows and diagonals (columns and diagonals when the rows are interchanged and the other diagonals, columns and rows when parallel diagonals are interchanged). Note that the transposing of two pairs of the (say upward) diagonals is done in such a manner as to make no change in the elements composing the other (downward) diagonals.

Such a square will be pandiagonal if C, R, c, r, $(Cr - cR)$, $(C \pm R)$ and $(c \pm r)$ are all prime to n. Also, it will be possible to make the desired interchanges if $(C^2 + R^2)$ and $(c^2 + r^2)$ are divisible by n. For example, let $C = 2$, $R = 3$, $c = 3$, $r = 2$ and $n = 13$. Now we have: $C + R = 5$, $C - R = -1$, $c + r = 5$, $c - r = 1$ and $(Cr - cR) = -5$, all of which are prime to 13. Also we have $(C^2 + R^2) = 13$ and $(c^2 + r^2)$ equal to 13, as required.

Figure 61 is the result obtained by following the normal procedure. Examination will show that each row, column and diagonal contains each capital letter and each lower-case letter once, and only once. It is thus a simple matter to assign numerical values so as to construct a normal pandiagonal magic square of the thirteenth order.

To determine which columns to interchange is quite simple. First select at random some cell to use as an arbitrary center, say column 8 and row 6 containing $(C + b)$ and draw two diagonals through it. Select, again at random, a pair of columns equidistant from the cell selected for the center, say columns 5 and 11. Note the letters in the cells located by the intersection of these columns and the diagonals previously drawn, namely, $(E + d)$, $(A + m)$, $(M + e)$ and $(F + l)$.

Now look for A and E on the opposite diagonal. You will find them in columns 6 and 10 (which are also symmetrically placed about our center cell). Note also that M and F also lie on the intersection of these columns with the two diagonals. In other words, if we interchange columns 5 and 11 and columns 6 and 10 we will find that the capital letters on the two diagonals (even though their positions relative to each other have been changed) remain the same. Further, a check on the lower-case letters will show that a similar situation exists here. By virtue of the cyclical construction of the square this will hold for all the diagonals. Figure 62 shows Figure 61 with these columns interchanged.

Careful inspection of Figure 62 will show that each row, column and diagonal contains each capital letter and each lower-case letter once, and only once, and that, while the letters appear in a cyclical

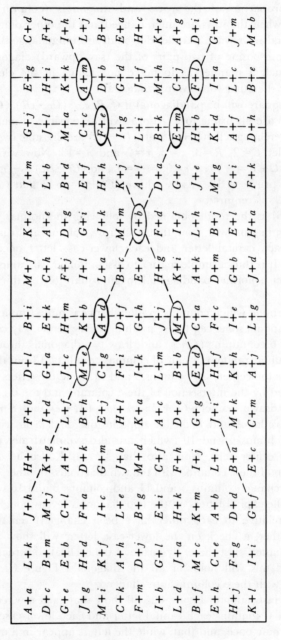

Fig. 61.

$A+a$	$L+k$	$J+h$	$H+e$	$F+b$	$E+g$	$G+j$	$M+f$	$K+c$	$I+m$	$B+i$	$D+l$	$C+d$
$D+c$	$B+m$	$M+j$	$K+g$	$I+d$	$H+i$	$J+l$	$C+h$	$A+e$	$L+b$	$E+k$	$G+a$	$F+f$
$G+e$	$E+b$	$C+l$	$A+i$	$L+f$	$K+k$	$M+a$	$F+j$	$D+g$	$B+d$	$H+m$	$J+c$	$I+h$
$J+g$	$H+d$	$F+a$	$D+k$	$B+h$	$A+m$	$C+c$	$I+l$	$G+i$	$E+f$	$K+b$	$M+e$	$L+j$
$M+i$	$K+f$	$I+c$	$G+m$	$E+j$	$D+b$	$F+e$	$L+a$	$J+k$	$H+h$	$A+d$	$C+g$	$B+l$
$C+k$	$A+h$	$L+e$	$J+b$	$H+l$	$G+d$	$I+g$	$B+c$	$M+m$	$K+j$	$D+f$	$F+i$	$E+a$
$F+m$	$D+j$	$B+g$	$M+d$	$K+a$	$J+f$	$L+i$	$E+e$	$C+b$	$A+l$	$G+h$	$I+k$	$H+c$
$I+b$	$G+l$	$E+i$	$C+f$	$A+c$	$M+h$	$B+k$	$H+g$	$F+d$	$D+a$	$J+j$	$L+m$	$K+e$
$L+d$	$J+a$	$H+k$	$F+h$	$D+e$	$C+j$	$E+m$	$K+i$	$I+f$	$G+c$	$M+l$	$B+b$	$A+g$
$B+f$	$M+c$	$K+m$	$I+j$	$G+g$	$F+l$	$H+b$	$A+k$	$L+h$	$J+e$	$C+a$	$E+d$	$D+i$
$E+h$	$C+e$	$A+b$	$L+l$	$J+i$	$I+a$	$K+d$	$D+m$	$B+j$	$M+g$	$F+c$	$H+f$	$G+k$
$H+j$	$F+g$	$D+d$	$B+a$	$M+k$	$L+c$	$A+f$	$G+b$	$E+l$	$C+i$	$I+e$	$K+h$	$J+m$
$K+l$	$I+i$	$G+f$	$E+c$	$C+m$	$B+e$	$D+h$	$J+d$	$H+a$	$F+k$	$L+g$	$A+j$	$M+b$

Fig. 62.   A noncyclical regular thirteenth-order pandiagonal magic square.

order in the columns, they do not in the rows and diagonals. It follows that Figure 62 is regular and pandiagonal but not cyclical.

It should perhaps be noted that the smallest-order magic square which can be constructed of this type is the thirteenth-order. While the requirements for C, R, c and r can be met for a fifth-order square, and the square actually constructed and the two pairs of columns interchanged, still retaining the pandiagonal property, the final square will (due to its small size) still be cyclical.

PART III

ENUMERATION OF MAGIC SQUARES

16

INTRODUCTORY

As previously noted, the history of magic squares is an ancient one. In recent years much progress has been made in methods of construction of normal magic squares. As described in Part I, we are now in a position to construct such squares of any order above two in a perfectly straightforward manner. Also, as we saw in Part II, we are frequently in a position to predict—in advance of actual construction and mathematical verification—magical properties over and above the normal ones.

There remains a third avenue of investigation which has intrigued the experts and amateurs over the years, namely, that of enumeration. How many magic squares are possible of any given order or classification? Little real progress has been made in this direction other than in a few rather limited cases.

While the census for the third order is complete, and that of the fourth-order magic squares is believed to be, that of the fifth and higher orders is wide open and likely to remain so!

17

THIRD-ORDER MAGIC SQUARES

The basic theory of third-order magic squares is complete and, since it is quite short, we will include it in its entirety. Let us assume that Figure 63a is magic.

a	b	c
d	e	f
g	h	i

Fig. 63a.

It follows that the following equations will hold:

$$a + e + i = 3m \qquad c + e + g = 3m$$

$$b + e + h = 3m \qquad d + e + f = 3m,$$

where m is the average value of the numbers forming the square, that is:

$$m = (a+b+c+d+e+f+g+h+i)/9.$$

If we add these four equations together we get

$$(a+b+c+d+e+f+g+h+i) + 3e = 12m$$

or

$$9m + 3e = 12m$$

or

$$e = m$$

and hence

$$a+i = g+c = 2m.$$

Let $a = m - p$ and $g = m - q$. Then $i = m + p$ and $c = m + q$. Now let us substitute these values in Figure 63a, getting Figure 63b.

$m-p$	b	$m+q$
d	m	f
$m-q$	h	$m+p$

Fig. 63b.

For this square to be magic it is necessary that:

$$(m-p) + b + (m+q) = 3m$$

In other words, b must equal $(m+p-q)$. In a similar manner it can be shown that d must equal $(m+p+q)$, h must equal $(m-p+q)$ and f must equal $(m-p-q)$. Substituting these values gives us our final square, Figure 63c.

$m-p$	$m+p-q$	$m+q$
$m+p+q$	m	$m-p-q$
$m-q$	$m-p+q$	$m+p$

Fig. 63c.

A quick check will show that Figure 63c will be magic regardless of the values selected for m, p and q. It will also show that a magic square of the third order must be composed of the following three arithmetical progressions (all with the same constant difference q):

$$
\begin{array}{ccc}
m-p-q & m-p & m-p+q \\
m-q & m & m+q \\
m+p-q & m+p & m+p+q.
\end{array}
$$

Notice further that corresponding terms of these three progressions must also form an arithmetical progression with the constant difference p.

It is thus evident that given any three arithmetical progressions (with the same constant difference) whose first three terms also form an arithmetical progression (with a different constant difference) you can form a third-order magic square and that, given any nine numbers which do not have this property, you will be unable to form such a square.

It is easy to divide the natural numbers 1 to 9, inclusive, into three such progressions (1-2-3, 4-5-6 and 7-8-9, and 1-4-7, 2-5-8 and 3-6-9) and hence we can form a normal magic square. We now have $m = 5$. If both p and q are even, all numbers will be odd and the square cannot be normal. If one is odd and the other even, then the square will consist of six even numbers and three odd numbers. Also note that p cannot equal q, for then we would have repetition, which is not allowed. It follows that if we want the square to be a normal one we must limit the values of p and q to some combination of 1, 3, -1 and -3. Substituting all possible combinations of these values in Figure 63c gives us Figure 64.

Notice that Figure 64 is actually a repeat of Figure 37. Any one of the four squares in either row is merely one of the other squares rotated through 90°, 180° or 270°; either one of the two squares in the same column is merely a mirror image of the other square. Hence these eight squares are equivalent to each other.

There is, therefore, only one essentially different magic square of the third order. Note also that, while this square is symmetrical, it is not pandiagonal.

It might be of interest to see how the above criteria can be used to construct Figure 2, a magic square of the third order using the smallest possible prime numbers (considering 1 to be a prime number). A little reflection will suggest trying 6 as the constant difference q. Examination of a table of primes gives us the following

8	1	6
3	5	7
4	9	2

$p = -3, q = 1$

4	3	8
9	5	1
2	7	6

$p = 1, q = 3$

2	9	4
7	5	3
6	1	8

$p = 3, q = -1$

6	7	2
1	5	9
8	3	4

$p = -1, q = -3$

4	9	2
3	5	7
8	1	6

$p = 1, q = -3$

2	7	6
9	5	1
4	3	8

$p = 3, q = 1$

6	1	8
7	5	3
2	9	4

$p = -1, q = 3$

8	3	4
1	5	9
6	7	2

$p = -3, q = -1$

Fig. 64.

arithmetic progressions with the constant difference 6:

1	7	13		17	23	29		61	67	73
5	11	17		31	37	43		67	73	79
7	13	19		41	47	53		91	97	103
11	17	23		47	53	59		and so on.		

Considering the first number in each series and investigating the various possible values of p, it will be found that 30 is the first that will work. This gives us 1-7-13, 31-37-43 and 61-67-73 as the smallest set of prime numbers meeting the above criteria. A similar check with $q = 12$ will give us 29-41-53, 59-71-83 and 89-101-113, which is, of course, not as good as the first set. Larger differences will require still larger prime numbers.

Using the above values, we get Figure 65 (which is identical with Figure 2), a third-order magic square composed of prime numbers so selected as to have the smallest possible magic constant, namely 111.

67	1	43
13	37	61
31	73	7

Fig. 65.   A third-order magic square formed of prime
numbers.

18

FOURTH-ORDER MAGIC SQUARES

In 1693 Frénicle published the then known magic squares of the fourth order—880 in all. His count has been verified from time to time. However, to date no one has been able to find that eight hundred and eighty-first fourth-order magic square!

Recently the authors, using the facilities of the Computer Center of Dickinson College, Carlisle, Pennsylvania, constructed a program designed to generate all possible fourth-order magic squares. The IBM 1130 Computer proceeded to generate all 880 of these squares and then stated, "COMPUTATION IS FINISHED!!!," thus confirming that no more are possible.

With the kind assistance of Mr. Michael O'Heeron, the Director of Academic Computing, the program was extended to indicate the index number and the type (as described below) of the squares and to print all the 880 possible squares in tabular form. As a matter of convenience to the reader who may be interested in constructing (or in identifying) such squares, this table has been reproduced in the Appendix.

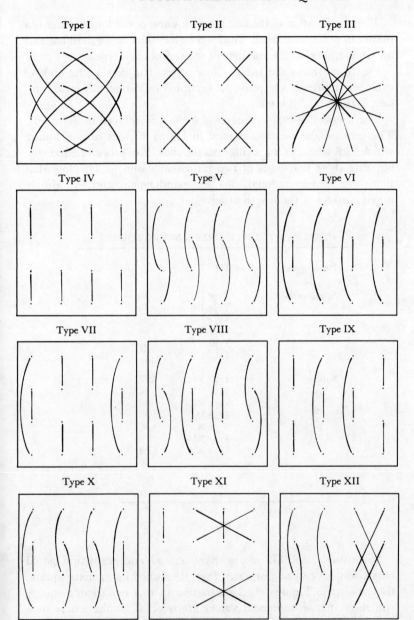

Fig. 66. The twelve basic types of fourth-order magic squares.

The classification of the 880 magic squares of the fourth order is a matter of preference. H. E. Dudeney* gives the one we shall use and an enumeration of the number of squares of each type.

Figure 66 shows the twelve basic types. The dots at the ends of each line represent the positions occupied by complementary numbers 1-16, 2-15, 3-14, etc.

The number of magic squares of each type is given in Figure 67. The term *semipandiagonal*, as used in Figure 67, describes a square which, in addition to being magic, has the property that the opposite short diagonals of two terms each sum to 34. Note that there are 48 pandiagonal, 384 semipandiagonal and 448 simple magic squares of the fourth order.

CENSUS OF FOURTH-ORDER MAGIC SQUARES

Pandiagonal	Type I		48
Semipandiagonal	Type II	48	
	Type III	48	
	Type IV	96	
	Type V	96	
	Type VI	96	
	Total		384
Simple	Type VI	208	
	Type VII	56	
	Type VIII	56	
	Type IX	56	
	Type X	56	
	Type XI	8	
	Type XII	8	
	Total		448
All Types	Total		880

Fig. 67.

Surprisingly, all 432 of the pandiagonal and semipandiagonal magic squares can be generated from the same intermediate square, that shown in Figure 68. Examination of this remarkable square will show that if numerical values are selected for the letters such

---

Amusements in Mathematics by H. E. Dudeney, Dover Publications, Inc., pp. 119–21.

that the resultant numbers forming the square are 1 to 16, inclusive, the square thus generated will be one of the types I to VI.

$A + a$	$B + b$	$C + c$	$D + d$
$D + c$	$C + d$	$B + a$	$A + b$
$B + d$	$A + c$	$D + b$	$C + a$
$C + b$	$D + a$	$A + d$	$B + c$

Fig. 68.

Which type it is will depend upon which one of the three possible combinations of capital letters ($A + B = C + D$, $A + C = B + D$ or $A + D = B + C$) is combined with which one of the three possible combinations of lower-case letters ($a + b = c + d$, $a + c = b + d$ or $a + d = b + c$).

Figure 69 lists the nine possible combinations and the type square that each combination will generate.

Capital Letters	Lower-case Letters	Type
$A + B = C + D$	$a + b = c + d$	IV
	$a + c = b + d$	III
	$a + d = b + c$	V
$A + C = B + D$	$a + b = c + d$	VI
	$a + c = b + d$	V
	$a + d = b + c$	II
$A + D = B + C$	$a + b = c + d$	I
	$a + c = b + d$	IV
	$a + d = b + c$	VI

Fig. 69.

Let us consider first the use of 0, 4, 8 and 12 for the capital letters and 1, 2, 3 and 4 for the lower-case letters. There are eight possible combinations of values for a, b, c and d which meet each of the above three requirements for lower-case letters, namely:

For $a+b=c+d$ we can use				For $a+c=b+d$ we can use				For $a+d=b+c$ we can use			
a	b	c	d	a	b	c	d	a	b	c	d
1	4	2	3	1	2	4	3	1	2	3	4
1	4	3	2	1	3	4	2	1	3	2	4
2	3	1	4	2	1	3	4	2	1	4	3
2	3	4	1	2	4	3	1	2	4	1	3
3	2	1	4	3	1	2	4	3	1	4	2
3	2	4	1	3	4	2	1	3	4	1	2
4	1	2	3	4	2	1	3	4	2	3	1
4	1	3	2	4	3	1	2	4	3	2	1

Continuing now with the values of the capital letters, we find that the following combinations not only meet the three requirements for capital letters but that the 16 magic squares which any pair of combinations will generate, when combined with any one of the above eight sets of values for the lower-case letters, will be different (no one of them will be a rotation, reflection or combination of rotation and reflection, of any other one). In other words, using the following combinations of values 0, 4, 8 and 12 for the capital letters, with the above combinations of values for the lower-case letters, we can generate 16 different magic squares of Types I, II and III, and 32 different magic squares of Types IV, V and VI (16 for each of the two possible combinations).

For $A+B=C+D$ we can use				For $A+C=B+D$ we can use				For $A+D=B+C$ we can use			
A	B	C	D	A	B	C	D	A	B	C	D
0	12	4	8	0	4	12	8	0	4	8	12
0	12	8	4	0	8	12	4	0	8	4	12

As an example, suppose you wish to generate a magic square of Type VI. Referring to Figure 68, you see that you can use either

$A + C = B + D$ combined with $a + b = c + d$, or $A + D = B + C$ combined with $a + d = b + c$, to get a magic square of Type VI. As we have just seen, there are 16 combinations of values which will meet each of these two combinations, 32 combinations in all. Suppose we select $A = 0$, $B = 4$, $C = 8$, $D = 12$, $a = 1$, $b = 3$, $c = 2$ and $d = 4$ and substitute these values in Figure 68, getting Figure 70a. Adding the two numbers in the same cell together gives Figure 70b—a semipandiagonal magic square of Type VI as predicted.

0 + 1	4 + 3	8 + 2	12 + 4
12 + 2	8 + 4	4 + 1	0 + 3
4 + 4	0 + 2	12 + 3	8 + 1
8 + 3	12 + 1	0 + 4	4 + 2

Fig. 70a.

1	7	10	16
14	12	5	3
8	2	15	9
11	13	4	6

Fig. 70b.

Note that the fact that it is the first and fourth numbers in each row, rather than the first and fourth numbers in each column, which are complementary does not keep this square from being Type VI. In fact, it is just this "directional property" of Types IV, V and VI (the semipandiagonal portion thereof) that gives us twice as many different magic squares as in the case of Types I, II and III.

Consider next the combination of 0, 2, 8 and 10 for the capital letters and 1, 2, 5 and 6 for the lower-case letters. Here again there are eight possible combinations of values which will meet each of the three basic requirements for the lower-case letters. Also we find that the following combinations of values for the capital letters not only meet the three basic requirements for the capital letters but that the 16 magic squares which any pair of combinations will generate, when combined with any of the eight sets of values for the lower-case letters, will be different (different not only from each other but from the 144 magic squares previously generated by using combinations of 0, 4, 8 and 12 with 1, 2, 3 and 4):

For $A+B=C+D$ we can use				For $A+C=B+D$ we can use				For $A+D=B+C$ we can use			
A	B	C	D	A	B	C	D	A	B	C	D
0	10	2	8	0	2	10	8	0	2	8	10
0	10	8	2	0	8	10	2	0	8	2	10

A similar situation exists when we use the following combinations of 0, 2, 4 and 6 for the capital letters in combination with the eight possible combinations of 1, 2, 9 and 10 for the lower-case letters:

For $A+B=C+D$ we can use				For $A+C=B+D$ we can use				For $A+D=B+C$ we can use			
A	B	C	D	A	B	C	D	A	B	C	D
0	6	2	4	0	2	6	4	0	2	4	6
0	6	4	2	0	4	6	2	0	4	2	6

Thus we see that we can construct all 432 of the fourth-order magic squares which are pandiagonal or semipandiagonal by simply inserting the above values in Figure 68. It should be noted that these combinations of values are not unique. That is to say, there are other combinations which could have been selected which would have the same property. On the other hand, once you have selected your set of values to generate the 432 pandiagonal and semipandiagonal magic squares, you will find that any other combination that will generate a normal magic square will result merely in generating one that is a duplicate of one of those previously generated.

Returning for a moment to Figure 67, we note that, in addition to the 96 semipandiagonal magic squares of Type VI, there are 208 simple magic squares of this type. While it is possible to construct intermediate squares that will permit these squares to be constructed directly, we shall take advantage of the technique of transformations to simplify our task.

Let Figure 71a be any magic square of Type VI. In this, and succeeding figures, $-a$ is the complement of a, $-b$ the complement of b, and so on. Since we are assuming that this square is a magic square of Type VI, it follows that:

$$a+c+e+g=b+d+f+h=a+d+(-f)+(-g)=34.$$

Now form Figure 71b by inverting and reversing the two center columns of Figure 71a. It is also a magic square of Type VI. The columns and main diagonals still contain the same numbers and the rows contain two pairs of complementary numbers.

As a next step form Figures 71c and 71d by interchanging the complementary pair in the second and third columns of the first row with the corresponding pair in the fourth row of Figures 71a and 71b, respectively. Now form Figures 71e, 71f, 71g and 71h by interchanging the complementary pair in the first and fourth columns of the second row with the corresponding pair in the third row of Figures 71a, 71b, 71c and 71d, respectively. Note that here again the interchanges leave the columns and main diagonals intact and the rows still consisting of two pairs of complementary numbers.

a	b	$-b$	$-a$
c	d	$-d$	$-c$
e	f	$-f$	$-e$
g	h	$-h$	$-g$

a	$-h$	h	$-a$
c	$-f$	f	$-c$
e	$-d$	d	$-e$
g	$-b$	b	$-g$

Fig. 71a.   *Fig. 71b.*

a	h	$-h$	$-a$
c	d	$-d$	$-c$
e	f	$-f$	$-e$
g	b	$-b$	$-g$

a	$-b$	b	$-a$
c	$-f$	f	$-c$
e	$-d$	d	$-e$
g	$-h$	h	$-g$

a	b	$-b$	$-a$
e	d	$-d$	$-e$
c	f	$-f$	$-c$
g	h	$-h$	$-g$

Fig. 71c.   *Fig. 71d.*   *Fig. 71e.*

a	$-h$	h	$-a$
e	$-f$	f	$-e$
c	$-d$	d	$-c$
g	$-b$	b	$-g$

Fig. 71f.

a	h	$-h$	$-a$
e	d	$-d$	$-e$
c	f	$-f$	$-c$
g	b	$-b$	$-g$

Fig. 71g.

a	$-b$	b	$-a$
e	$-f$	f	$-e$
c	$-d$	d	$-c$
g	$-h$	h	$-g$

Fig. 71h.

Examination of Figure 71 will show that if

$$b + c + (-e) + (-h) = 34,$$

then Figures 71a, 71b, 71g and 71h will be semipandiagonal and the remaining figures will be simple magic squares. Similarly if

$$c + h + (-e) + (-b) = 34,$$

then Figures 71c, 71d, 71e and 71f will be semipandiagonal and the remaining figures will be simple magic squares. Finally note that if neither of these two equations is true, all figures will be simple magic squares.

When any of the figures are semipandiagonal they must be duplicates of one of the 96 semipandiagonal magic squares of Type VI previously generated. It follows that when Figure 71a is one of the known semipandiagonal magic squares of Type VI, Figure 71c is a simple magic square of Type VI different from any we have generated so far. Hence, by using the 96 known semipandiagonal magic squares that we have, and applying the transformation of Figure 71c, we get 96 of the 208 simple magic squares of Type VI for which we are looking. This leaves us 112 to find.

If we attempt to reduce this figure still further by using the transformations of Figures 71d, 71e or 71f, we find that we get duplications of the squares generated by the use of Figure 71c.

However, we have just seen in our discussion of Figure 71 that, if we can find one new simple magic square of Type VI, we can by the transformations of Figure 71 generate seven more—provided $b + c + (-e) + (-h) \neq 34$ and $c + h + (-e) + (-b) \neq 34$. This means that all we really have to find is $112/8 = 14$ new simple magic squares of Type VI which do not duplicate any of the 96 we generated from the 96 semipandiagonal magic squares.

Fortunately this is not too difficult. We do, however, need three intermediate squares, Figures 72a, 72b and 72c.

$A+a$	$A+b$	$D+c$	$D+d$
$C+d$	$D+b$	$A+c$	$B+a$
$D+a$	$B+c$	$C+b$	$A+d$
$B+d$	$C+c$	$B+b$	$C+a$

Fig. 72a.

Figure 72a will be a simple magic square of Type VI meeting the above requirements, provided:

$$A + D = B + C, \qquad a + d = b + c,$$
$$C = B + k, \text{ where } k \text{ is any integer, and } a + b + k = c + d.$$

There are eight ways these conditions can be met and still have the squares generated from Figure 72a be different from each other and also different from our previous squares, namely:

A	B	C	D	a	b	c	d	k
0	4	8	12	1	2	3	4	+4
0	4	8	12	2	1	4	3	+4
0	8	4	12	3	4	1	2	−4
0	8	4	12	4	3	2	1	−4
0	2	4	6	1	9	2	10	+2
0	2	4	6	9	1	10	2	+2
0	4	2	6	2	10	1	9	−2
0	4	2	6	10	2	9	1	−2

Figure 72b will be a simple magic square meeting the above requirements, provided:

$$A + D = B + C, \qquad a + d = b + c,$$
$$C = B + k, \text{ where } k \text{ is any integer, and } 2a + k = b + d.$$

$A + a$	$A + c$	$D + b$	$D + d$
$C + d$	$D + a$	$A + d$	$B + a$
$D + c$	$B + d$	$C + a$	$A + b$
$B + b$	$C + b$	$B + c$	$C + c$

Fig. 72b.

There are three ways these conditions can be met and still have the squares generated by Figure 72b be different from our previous squares, namely:

A	B	C	D	a	b	c	d	k
0	4	8	12	1	2	3	4	$+4$
0	8	2	10	6	5	2	1	-6
2	0	6	4	2	1	10	9	$+6$

$A + b$	$A + a$	$D + d$	$D + c$
$C + d$	$D + b$	$A + c$	$B + a$
$D + a$	$B + d$	$C + a$	$A + d$
$B + c$	$C + c$	$B + b$	$C + b$

Fig. 72c.

Similarly, the requirements that must be met for the squares generated from Figure 72c to be simple magic squares of Type VI are:

$$A + D = B + C, \qquad a + d = b + c,$$

$C = B + k$, where k is any integer, and $2b + k = c + d$.

There are also three ways these conditions can be met and still have the squares generated by Figure 72c be different from our previous squares, namely:

A	B	C	D	a	b	c	d	k
0	8	4	12	2	4	1	3	−4
0	2	8	10	5	1	6	2	+6
2	6	0	4	1	9	2	10	−6

We now have 14 simple magic squares of Type VI which, when transposed in accordance with Figure 71 and added to the 96 squares we obtained previously (by transposing the 96 semipandiagonal Type VI magic squares), will give the 208 simple magic squares of Type VI required.

Let us now turn to Type VII. Here two intermediate squares will suffice, Figures 73a and 73b.

$A+a$	$B+b$	$C+c$	$D+d$
$C+b$	$C+d$	$B+a$	$B+c$
$B+d$	$D+b$	$A+c$	$C+a$
$D+c$	$A+d$	$D+a$	$A+b$

Fig. 73a.

$A+a$	$B+b$	$C+a$	$D+b$
$B+c$	$D+a$	$A+b$	$C+d$
$D+d$	$C+c$	$B+d$	$A+c$
$C+b$	$A+d$	$D+c$	$B+a$

Fig. 73b.

Figure 73a will be a simple magic square of Type VII provided:

$$A + D = B + C, \qquad a + c = b + d, \qquad \text{and} \quad 2A = B + D.$$

There are 24 ways these conditions can be met and still have the squares that are generated from Figure 73a be different, namely:

A	B	C	D	Combined with the eight possible combinations of the following values for a, b, c and d, so arranged as to make $a + c = b + d$
2	1	4	3	combined with 0, 4, 8, and 12,
2	0	6	4	combined with 1, 2, 9, and 10,
4	0	12	8	combined with 1, 2, 3, and 4.

Figure 73b will be a simple magic square of Type VII provided:

$$A + C = B + D, \qquad a + b = c + d,$$

$C = B + k$, where k is any integer, and $2b + k = a + d$.

There are four ways these conditions can be met and still have the squares that are generated from Figure 73b be different from those generated previously, namely:

A	B	C	D	a	b	c	d	k
0	8	12	4	4	1	3	2	+4
12	4	0	8	1	4	2	3	−4
1	2	6	5	8	2	10	0	+4
6	5	1	2	2	8	0	10	−4

These two squares thus enable us to construct 28 different simple magic squares of Type VII. We now resort to the same transposition device to double this number and thus generate all 56 of the Type VII magic squares. Assume Figure 74a is a magic square of Type VII and interchange the complementary pair in the second and third rows of the first column with the corresponding pair in the fourth column to get Figure 74b. Clearly this will result in generating a new simple magic square of Type VII. It is thus easy to use this transposition to generate 28 new Type VII simple magic squares from the known set of 28 squares.

a	b	c	d
e	$-b$	$-c$	f
$-e$	g	h	$-f$
$-a$	$-g$	$-h$	$-d$

Fig. 74a.

a	b	c	d
f	$-b$	$-c$	e
$-f$	g	h	$-e$
$-a$	$-g$	$-h$	$-d$

Fig. 74b.

Having thus generated all 56 of the Type VII simple magic squares of the fourth order, it remains to generate the 56 squares of Types VIII, IX and X. These can be generated in a fashion similar to that employed in generating the 56 Type VII squares but the simplest way is to permute the rows and columns of the Type VII squares. If we permute the rows and columns of a Type VII square in the order 3-1-4-2, we will generate a Type VIII square; if we use the order 2-1-4-3, a Type IX square; and if we use the order 1-3-2-4, a Type X square. For example, suppose we are given the Type VII simple magic square in Figure 75a. Figure 75b is the Type VIII square derived from it by permuting the rows and columns in the order 3-1-4-2. In a like manner Figure 75c is the Type IX simple magic square derived from Figure 75a by permuting the rows and columns in the order 2-1-4-3. Finally, the Type X square in Figure 75d can be derived from Figure 75a by using the permutation 1-3-2-4.

2	5	16	11
8	12	1	13
9	7	14	4
15	10	3	6

Fig. 75a.   Type VII

14	9	4	7
16	2	11	5
3	15	6	10
1	8	13	12

Fig. 75b.   Type VIII

12	8	13	1
5	2	11	16
10	15	6	3
7	9	4	14

Fig. 75c.   Type IX

2	16	5	11
9	14	7	4
8	1	12	13
15	3	10	6

Fig. 75d.   Type X

It follows that, given the 56 simple magic squares of Type VII, it is a simple matter to generate the 56 simple magic squares of Types VIII, IX and X.

There remain only Types XI and XII. The eight simple magic squares of Type XI can be generated by substituting in Figure 76 the following eight combinations of values:

A	B	C	D	a	b	c	d	k
0	4	8	12	1	2	3	4	+4
0	4	8	12	2	1	4	3	+4
0	8	4	12	3	4	1	2	−4
0	8	4	12	4	3	2	1	−4
0	2	4	6	1	9	2	10	+2
0	2	4	6	9	1	10	2	+2
0	4	2	6	2	10	1	9	−2
0	4	2	6	10	2	9	1	−2

$A+d$	$B+a$	$D+d$	$C+a$
$D+a$	$B+d$	$A+a$	$C+d$
$A+c$	$C+b$	$D+c$	$B+b$
$D+b$	$C+c$	$A+b$	$B+c$

Fig. 76.

$A+d$	$D+d$	$B+a$	$C+a$
$A+c$	$D+c$	$C+b$	$B+b$
$D+a$	$A+a$	$B+d$	$C+d$
$D+b$	$A+b$	$C+c$	$B+c$

Fig. 77.

Notice that these values meet the obvious requirements that:

$$A + D = B + C, \qquad a + d = b + c,$$

$C = B + k$, where k is any integer, and $a + b + k = c + d$.

The same requirements hold for Figure 77, that is, if they are met the generated square will be Type XII. In other words, we can easily generate the eight simple magic squares of Type XII by substituting in Figure 77 the same values we used for Type XI in Figure 76. Or, for that matter, we can take the eight Type XI squares and, by permuting the rows and columns in the order 1-3-2-4, generate the eight Type XII simple magic squares directly without resorting to Figure 77.

For the benefit of the reader who might be inclined to undertake the task of constructing the complete set of 880 fourth-order magic squares, we will next discuss the method commonly used to index magic squares. Incidentally, we will need it in the next chapter. The explanation is based upon the one used by Frénicle as described by K. H. De Haas.*

Obviously, before any magic square can be indexed it is necessary to establish a standard, or normalized, position. Otherwise, we would have great difficulty in making sure that each magic square in our index was different—that no square was a rotation, reflection or combination of rotation and reflection, of any other square. The first requirement for a magic square in the standard position is: the cell occupied by the smallest of the numbers in the four corner cells must be in the upper left-hand corner. If this is not the case, we rotate the square until it is. The second, and only additional, requirement that must be met is: the number in the cell in the top row adjacent to the upper left-hand corner must be smaller than the number in the cell in the left-hand column adjacent to the upper left-hand corner cell. If it is we are finished. The square is in the standard position. If it is not, we must mirror the square about the downward main diagonal so that it will be. This may sound complicated, but it isn't, as the following example will demonstrate.

Suppose we desire to put Figure 75c into standard position. Our first step would be to rotate the square 90° counterclockwise, thus

---

*Frénicle's 880 Basic Magic Squares of 4 × 4 cells, Normalized, Indexed, and Inventoried, by K. H. De Haas, D. Van Sijn & Zonen, Holland.

putting the number 1 in the upper left-hand corner (see Figure 75e). Since 16 is not less than 13 the square still is not in the standard position. It is necessary to mirror it about the downward diagonal to get Figure 75f, which is in the standard position. It is, of course, equivalent to the original square (Figure 75c) since it was derived from it by a combination of rotation and reflection.

1	16	3	14
13	11	6	4
8	2	15	9
12	5	10	7

1	13	8	12
16	11	2	5
3	6	15	10
14	4	9	7

Fig. 75e.   *Fig. 75f.*

Note particularly that the above definition of standard position is applicable to magic squares of any order. This statement also applies to the actual indexing process.

Indexing is, of necessity, limited to a set of squares of one order—what this order is is immaterial, but all the squares in the set must be of the same order. It is essentially an ordering process. Given any two magic squares, the indexing process defines the method to be employed to determine which one shall be given a lower index number. Once all the squares forming the set have been ordered, it is a simple matter to number each consecutively, beginning with 1 for the square to be given the lowest index number of all.

The ordering process is simplicity itself. Examine the top rows of any two squares in the standard position. If they are identical, check the next-to-top rows. If these are identical, the next, and so on. Eventually, if the squares are different, you will reach a pair of rows that differ in some cell. Take the first cell that is different, counting from left to right. The square that has the lower of the two numbers occupying this cell is given the lower index number.

19

FIFTH-ORDER
PANDIAGONAL MAGIC SQUARES

It is well known that by using certain substitutions and permutations the 3600 possible fifth-order pandiagonal magic squares can be constructed from four basic squares. It is not so well known that, as was stated in Chapter 9, they can be constructed from one basic square, namely, that shown in Figure 78. This square is identical with that shown in Figure 36d, only the letters have been shifted so as to make the first row simpler.

It remains to show how any particular one of these 3600 squares can be generated without going through the very cumbersome process of generating all 28,880 pandiagonal magic squares, placing them in standard position, throwing out the duplicates, ranking the remaining 3600 and, finally, assigning index numbers according to Frénicle's method (as described at the end of the last chapter). Of course, it follows that if we can generate any particular square, we can generate any set of squares (up to the entire set of 3600) that we desire.

$A + a$	$B + b$	$C + c$	$D + d$	$E + e$
$C + d$	$D + e$	$E + a$	$A + b$	$B + c$
$E + b$	$A + c$	$B + d$	$C + e$	$D + a$
$B + e$	$C + a$	$D + b$	$E + c$	$A + d$
$D + c$	$E + d$	$A + e$	$B + a$	$C + b$

Fig. 78.   Basic Square Number 1. Fifth-order basic
pandiagonal magic square.

As we have seen, by using Figure 78 we can generate all 3600
normal fifth-order pandiagonal magic squares if we substitute 0, 5,
10, 15 and 20 for the capital letters and 1, 2, 3, 4 and 5 for the
lower-case letters and also substitute 1, 2, 3, 4 and 5 for the capital
letters and 0, 5, 10, 15 and 20 for the lower-case letters.

Since it will simplify the presentation of the following method, we
will construct a second basic square by interchanging the capital
and lower-case letters in Figure 78 (see Figure 79).

$A + a$	$B + b$	$C + c$	$D + d$	$E + e$
$D + c$	$E + d$	$A + e$	$B + a$	$C + b$
$B + e$	$C + a$	$D + b$	$E + c$	$A + d$
$E + b$	$A + c$	$B + d$	$C + e$	$D + a$
$C + d$	$D + e$	$E + a$	$A + b$	$B + c$

Fig. 79.   Basic Square Number 2. Fifth-order basic
pandiagonal magic square.

A simple check will show that if you pick any set of values, say
$A = 2, B = 3, C = 1, D = 5, E = 4, a = 0, b = 10, c = 15, d = 5$ and $e = 20$,
and use these values in Figure 78 to generate a particular magic
square and then use the reverse values ($A = 0, B = 10, C = 15, D = 5$,
$E = 20, a = 2, b = 3, c = 1, d = 5$ and $e = 4$) in Figure 79, you will
generate the identical magic square.

In other words, while it is possible to generate all 3600 normal fifth-order pandiagonal magic squares from Figure 78, by using both Figures 78 and 79 it is possible to generate all 3600 squares by using only 0, 5, 10, 15 and 20 for the capital letters and 1, 2, 3, 4 and 5 for the lower-case letters.

As a matter of possible interest note that Figure 79 can be also obtained from Figure 78 by permuting the rows in the order 1-5-4-3-2.

We are now in a position, with the aid of the tables at the end of this chapter, to generate any specified fifth-order pandiagonal magic square. Obviously, any square generated by the use of Figure 79 could have been generated by using Figure 78 and reversing the values used between the capital and lower-case letters.

STEP I.   Enter the left-hand column of Table 5 with Frénicle's index number of the desired square. This will enable you to determine the values to use for A, B, a and b, the class letter of the square and the code number from the formula in the right-hand column.

STEP II.   Refer next to Table 6-P, 6-Q, 6-R or 6-S, as appropriate, to determine the values of C, D, E, c, d and e. The table to be used is, of course, the table for the class to which the desired square belongs. In order to avoid needless repetition in these tables, the values of c, d and e are given in terms of f, g and h, where the values to be assigned to f, g and h are the values remaining in the set 1, 2, 3, 4 and 5 after those assigned to a and b have been deleted. In addition, these values are assigned so that f is less than g and g is less than h.

For example, suppose you wish to generate square number 1457 (in Frénicle's index). Referring to Table 5, we find that 1457 lies in the range 1441 to 1488 and that for squares in this range $A = 0$, $B = 10$, $a = 3$ and $b = 1$, thus making $f = 2$ less than $g = 4$ less than $h = 5$. We also note that it is in Class Q with code number $q = n - 1440 = 1457 - 1440 = 17$.

Turning now to Table 6-Q for Class Q, we find that for code number $q = 17$ we should use Basic Square Number 1 with $C = 15$, $D = 20$, $E = 5$, $c = f = 2$, $d = h = 5$ and $e = g = 4$. To recapitulate, we now have $A = 0$, $B = 10$, $C = 15$, $D = 20$, $E = 5$, $a = 3$, $b = 1$, $c = 2$, $d = 5$ and $e = 4$. Substituting these values in Basic Square Number 1, Figure 78, gives us Figures 80a and 80b.

0+3	10+1	15+2	20+5	5+4
15+5	20+4	5+3	0+1	10+2
5+1	0+2	10+5	15+4	20+3
10+4	15+3	20+1	5+2	0+5
20+2	5+5	0+4	10+3	15+1

Fig. 80a.

3	11	17	25	9
20	24	8	1	12
6	2	15	19	23
14	18	21	7	5
22	10	4	13	16

Fig. 80b.   Pandiagonal square 1457.

The next problem of interest is—given a normal fifth-order pandiagonal magic square, how many pandiagonal magic squares can be formed (or derived) from it by cyclical permutations, transformations of rows and/or columns and interchanges of rows and columns with the diagonals? Since the top row can be permuted to the bottom of the square without disturbing its magical properties, it follows that five (including the original) different pandiagonal magic squares can be generated in this manner. The columns possess this same property, thus giving us a total of five squares for each of the previous five for a total of 25 in all. In other words, by cyclically permuting the rows and columns, we can place any one of the 25 numbers forming the square in the upper left-hand corner of the square. It remains only to examine the remaining possible transformations to see which ones, if any, will generate a new square and, at the same time, retain the pandiagonal magic properties of the original square.

Investigation shows that there are two transformations with this property. We can permute the rows *and* columns in the order 1-3-5-2-4. Naturally any cyclical rearrangement of these numbers, or reversal thereof, will also work but these are not essentially different transformations. We can also interchange the rows and columns with the diagonals. It follows, therefore, that we have the original square, the 1-3-5-2-4 transformation thereof, the interchange for the rows and columns with the diagonals for each of these squares, and the 24 additional squares we can get from each of these four squares by cyclical permutations, for a grand total of 100 pandiagonal magic squares (including, of course, the original square) that can be obtained from any pandiagonal magic square by means of transformations and permutations.

We thus see that our 3600 fifth-order pandiagonal magic squares reduce to $3600/100 = 36$ fifth-order pandiagonal magic squares which are essentially different from each other in that no one can be obtained by any combination of the 1-3-5-2-4 transformation, the interchange of the rows and columns with the diagonals, and/or cyclical permutations.

How can we identify which square of any different set of 100 is the "essentially different" one? It is somewhat like the problem that faced us when we wanted to know which one of eight equivalent magic squares was *the* square. For convenience we define a magic square as being *essentially different* when:

(a) the number in the upper left-hand corner cell is 1,

(b) the number in the cell next to 1 in the top row is less than any other number in the top row, in the left-hand column or in either diagonal containing the number 1, and

(c) the number in the second row of the left-hand column is less than the number in the last row in the left-hand column.

As an example look at Figure 81a, which is pandiagonal square #1233. Clearly, it is not an essentially different magic square. The first step in finding the essentially different square to which it is equivalent is to cyclically permute the rows and columns to put the number 1 in the upper left-hand corner. This gives us Figure 81b. We now see that 7 is the smallest number in the top row, the left-hand column and both of the diagonals through the number 1.

It follows that we must interchange the diagonals with the rows and columns in order to bring it into the top row. While it is possible to interchange the diagonals with the rows and columns in such a manner as to bring the number 7 into the left-hand column, there is no advantage in this as we would then have to go through the extra step of mirroring about the downward main diagonal in order to get it into the top row where we want it. This gives us Figure 81c.

We notice that the number 7 is in the top row as we desire, but it is in the third column. We now apply the transformation 1-3-5-2-4 in order to bring it into the cell adjacent to the number 1. If the number 7 had been in the fourth column of the top row we would have used the transformation 1-4-2-5-3 and if it had been in the last column we would have used 1-5-4-3-2 (applied to the columns only) to transform the square into an equivalent square with 7 adjacent to 1. This gives us Figure 81d.

3	7	14	16	25
11	20	23	2	9
22	4	6	15	18
10	13	17	24	1
19	21	5	8	12

Fig. 81a.   Pandiagonal magic square #1233.

1	10	13	17	24
12	19	21	5	8
25	3	7	14	16
9	11	20	23	2
18	22	4	6	15

Fig. 81b.   Fig. 81a permuted to bring number 1 to the left-hand corner.

1	19	7	23	15
8	25	11	4	17
14	2	18	10	21
20	6	24	12	3
22	13	5	16	9

Fig. 81c.   Fig. 81b with diagonals interchanged with rows and columns.

1	7	15	19	23
14	18	21	2	10
22	5	9	13	16
8	11	17	25	4
20	24	3	6	12

Fig. 81d.   Fig. 81c with transformation 1-3-5-2-4 applied to
rows and columns.

Note that Figure 81d has the number 7, the lowest number in the
top row, the left-hand column and in the diagonals through the
number 1—other than the number 1 itself—in the top row in the
cell adjacent to the upper left-hand corner, as required. A check
shows that it also meets the third requirement in that 14 is less than
20. It is, therefore, the essentially different square from which
square #1233 can be generated—in fact, by the reverse of the
process we just used to get it. (If the number in the second row of
the left-hand column had not been less than the number in the last
row in the left-hand column we would have had to permute the
rows only in the order 1-5-4-3-2 to get our final result.)

If we compare Figure 81d with Figures 78 and 79 we see that
$A = 0, B = 5, C = 10, D = 15, E = 20, a = 1, b = 2, c = 5, d = 4$ and $e = 3$.
Since $C + d = 10 + 4 = 14$, we know that this square was generated
from Figure 78, Basic Square Number 1. Furthermore, since $a = 1$
and $b = 2$, it follows that $f = 3 = e$, $g = 4 = d$ and $h = 5 = c$.

Using the above data and referring to Table 5, we see that Figure
81d must belong to Class P and have a code number $p = n - 0$. Our
next step is to refer to Table 6-P. Here we see that the required
values of $C = 10$, $D = 15$ and $E = 20$ indicate that the code number
will be somewhere between 1 and 4, 9 and 12 or 17 and 20. But we
know that $e = 3 = f$, $d = 4 = g$ and $c = 5 = h$ and that Basic Square
Number 1 was used. This gives us code number $p = 19$. Finally, from
the formula from Table 5, $p = n - 0$, we find that n, the Frénicle
index number of Figure 81d, is 19.

We thus see that square #1233 is one of the 100 different
pandiagonal magic squares that can be derived from the essentially
different magic square #19.

The question naturally arises, can all 36 essentially different

ENUMERATION OF MAGIC SQUARES

fifth-order pandiagonal magic squares be generated from a table similar to Tables 5 and 6 without any duplication? The answer is yes. Table 7 lists the squares and gives the essential data in one convenient place.

TABLE 5

Frénicle's Index Number	A	B	a	b	Class Letter	Code Number	Frénicle's Index Number	A	B	a	b	Class Letter	Code Number
n = 1–72	0	5	1	2	P	$p = n - 0$	2017–2064	0	10	4	1	Q	$q = n - 2016$
73–144	0	5	1	3	P	$p = n - 72$	2065–2112	0	10	4	2	Q	$q = n - 2064$
145–216	0	5	1	4	P	$p = n - 144$	2113–2160	0	10	4	3	Q	$q = n - 2112$
217–288	0	5	1	5	P	$p = n - 216$	2161–2208	0	10	4	5	Q	$q = n - 2160$
289–336	0	10	1	2	Q	$q = n - 288$	2209–2232	0	15	4	1	R	$r = n - 2208$
337–384	0	10	1	3	Q	$q = n - 336$	2233–2256	0	15	4	2	R	$r = n - 2232$
385–432	0	10	1	4	Q	$q = n - 384$	2257–2280	0	15	4	3	R	$r = n - 2256$
433–480	0	10	1	5	Q	$q = n - 432$	2281–2304	0	15	4	5	R	$r = n - 2280$
481–504	0	15	1	2	R	$r = n - 480$							
505–528	0	15	1	3	R	$r = n - 504$	2305–2376	0	5	5	1	P	$p = n - 2304$
529–552	0	15	1	4	R	$r = n - 528$	2377–2448	0	5	5	2	P	$p = n - 2376$
553–576	0	15	1	5	R	$r = n - 552$	2449–2520	0	5	5	3	P	$p = n - 2448$
							2521–2592	0	5	5	4	P	$p = n - 2520$
							2593–2640	0	10	5	1	Q	$q = n - 2592$
577–648	0	5	2	1	P	$p = n - 576$	2641–2688	0	10	5	2	Q	$q = n - 2640$
649–720	0	5	2	3	P	$p = n - 648$	2689–2736	0	10	5	3	Q	$q = n - 2688$
721–792	0	5	2	4	P	$p = n - 720$	2737–2784	0	10	5	4	Q	$q = n - 2736$
793–864	0	5	2	5	P	$p = n - 792$	2785–2808	0	15	5	1	R	$r = n - 2784$
865–912	0	10	2	1	Q	$q = n - 864$	2809–2832	0	15	5	2	R	$r = n - 2808$
913–960	0	10	2	3	Q	$q = n - 912$	2833–2856	0	15	5	3	R	$r = n - 2832$
961–1008	0	10	2	4	Q	$q = n - 960$	2857–2880	0	15	5	4	R	$r = n - 2856$
1009–1056	0	10	2	5	Q	$q = n - 1008$							
1057–1080	0	15	2	1	R	$r = n - 1056$	2881–2916	5	0	1	2	S	$s = n - 2880$
1081–1104	0	15	2	3	R	$r = n - 1080$	2917–2952	5	0	1	3	S	$s = n - 2916$
1105–1128	0	15	2	4	R	$r = n - 1104$	2953–2988	5	0	1	4	S	$s = n - 2952$
1129–1152	0	15	2	5	R	$r = n - 1128$	2989–3024	5	0	1	5	S	$s = n - 2988$
							3025–3060	5	0	2	1	S	$s = n - 3024$
1153–1224	0	5	3	1	P	$p = n - 1152$	3061–3096	5	0	2	3	S	$s = n - 3060$
1225–1296	0	5	3	2	P	$p = n - 1224$	3097–3132	5	0	2	4	S	$s = n - 3096$
1297–1368	0	5	3	4	P	$p = n - 1296$	3133–3168	5	0	2	5	S	$s = n - 3132$
1369–1440	0	5	3	5	P	$p = n - 1368$							
1441–1488	0	10	3	1	Q	$q = n - 1440$	3169–3204	5	0	3	1	S	$s = n - 3168$
1489–1536	0	10	3	2	Q	$q = n - 1488$	3205–3240	5	0	3	2	S	$s = n - 3204$
1537–1584	0	10	3	4	Q	$q = n - 1536$	3241–3276	5	0	3	4	S	$s = n - 3240$
1585–1632	0	10	3	5	Q	$q = n - 1584$	3277–3312	5	0	3	5	S	$s = n - 3276$
1633–1656	0	15	3	1	R	$r = n - 1632$							
1657–1680	0	15	3	2	R	$r = n - 1656$	3313–3348	5	0	4	1	S	$s = n - 3312$
1681–1704	0	15	3	4	R	$r = n - 1680$	3349–3384	5	0	4	2	S	$s = n - 3348$
1705–1728	0	15	3	5	R	$r = n - 1704$	3385–3420	5	0	4	3	S	$s = n - 3384$
							3421–3456	5	0	4	5	S	$s = n - 3420$
1729–1800	0	5	4	1	P	$p = n - 1728$	3457–3492	5	0	5	1	S	$s = n - 3456$
1801–1872	0	5	4	2	P	$p = n - 1800$	3493–3528	5	0	5	2	S	$s = n - 3492$
1873–1944	0	5	4	3	P	$p = n - 1872$	3529–3564	5	0	5	3	S	$s = n - 3528$
1945–2016	0	5	4	5	P	$p = n - 1944$	3565–3600	5	0	5	4	S	$s = n - 3564$

TABLE 6-P

Note 1: Enter table with the value of p determined by the formula in the right-hand column of Table 5.

Note 2: The values remaining in the set 1, 2, 3, 4 and 5, after a and b are assigned the values indicated in Table 5, shall be assigned to f, g and h so as to make f less than g less than h.

Code Number p	Basic Square to be used	C	D	E	c	d	e
1	1	10	15	20	f	g	h
2	2	"	"	"	"	"	"
3	1	"	"	"	"	h	g
4	2	"	"	"	"	"	"
5	1	"	20	15	"	g	h
6	2	"	"	"	"	"	"
7	1	"	"	"	"	h	g
8	2	"	"	"	"	"	"
9	1	"	15	20	g	f	h
10	2	"	"	"	"	"	"
11	1	"	"	"	"	h	f
12	2	"	"	"	"	"	"
13	1	"	20	15	"	f	h
14	2	"	"	"	"	"	"
15	1	"	"	"	"	h	f
16	2	"	"	·"	"	"	"
17	1	"	15	20	h	f	g
18	2	"	"	"	"	"	"
19	1	"	"	"	"	g	f
20	2	"	"	"	"	"	"
21	1	"	20	15	"	f	g
22	2	"	"	"	"	"	"
23	1	"	"	"	"	g	f
24	2	"	"	"	"	"	"
25	2	15	10	20	f	g	h
26	1	"	"	"	"	"	"
27	2	"	"	"	"	h	g
28	1	"	"	"	"	"	"
29	1	"	20	10	"	g	h
30	2	"	"	"	"	"	"
31	1	"	"	"	"	h	g
32	2	"	"	"	"	"	"
33	2	"	10	20	g	f	h
34	1	"	"	"	"	"	"
35	2	"	"	"	"	h	f
36	1	"	"	"	"	"	"
37	1	"	20	10	"	f	h
38	2	"	"	"	"	"	"
39	1	"	"	"	"	h	f
40	2	"	"	"	"	"	"
41	2	"	10	20	h	f	g
42	1	"	"	"	"	"	"
43	2	"	"	"	"	g	f
44	1	"	"	"	"	"	"
45	1	"	20	10	"	f	g
46	2	"	"	"	"	"	"
47	1	"	"	"	"	g	f
48	2	"	"	"	"	"	"
49	2	20	10	15	f	g	h
50	1	"	"	"	"	"	"
51	2	"	"	"	"	h	g
52	1	"	"	"	"	"	"
53	2	"	15	10	"	g	h
54	1	"	"	"	"	"	"
55	2	"	"	"	"	h	g
56	1	"	"	"	"	"	"
57	2	"	10	15	g	f	h
58	1	"	"	"	"	"	"
59	2	"	"	"	"	h	f
60	1	"	"	"	"	"	"
61	2	"	15	10	"	f	h
62	1	"	"	"	"	"	"
63	2	"	"	"	"	h	f
64	1	"	"	"	"	"	"
65	2	"	10	15	h	f	g
66	1	"	"	"	"	"	"
67	2	"	"	"	"	g	f
68	1	"	"	"	"	"	"
69	2	"	15	10	"	f	g
70	1	"	"	"	"	"	"
71	2	"	"	"	"	g	f
72	1	"	"	"	"	"	"

TABLE 6-Q

Note: Same as Table 6-P, except enter this table with the value q.

Code Number q	Basic Square to be used	C	D	E	c	d	e
1	2	5	15	20	f	g	h
2	2	"	"	"	"	h	g
3	2	"	20	15	"	g	h
4	2	"	"	"	"	h	g
5	2	"	15	20	g	f	h
6	2	"	"	"	"	h	f
7	2	"	20	15	"	f	h
8	2	"	"	"	"	h	f
9	2	"	15	20	h	f	g
10	2	"	"	"	"	g	f
11	2	"	20	15	"	f	g
12	2	"	"	"	"	g	f
13	1	15	5	20	f	g	h
14	1	"	"	"	"	h	g
15	1	"	20	5	"	g	h
16	2	"	"	"	"	"	"
17	1	"	"	"	"	h	g
18	2	"	"	"	"	"	"
19	1	"	5	20	g	f	h
20	1	"	"	"	"	h	f
21	1	"	20	5	"	f	h
22	2	"	"	"	"	"	"
23	1	"	"	"	"	h	f
24	2	"	"	"	"	"	"

Code Number q	Basic Square to be used	C	D	E	c	d	e
25	1	"	5	20	h	f	g
26	1	"	"	"	"	g	f
27	1	"	20	5	"	f	g
28	2	"	"	"	"	"	"
29	1	"	"	"	"	g	f
30	2	"	"	"	"	"	"
31	1	20	5	15	f	g	h
32	1	"	"	"	"	h	g
33	2	"	15	5	"	g	h
34	1	"	"	"	"	"	"
35	2	"	"	"	"	h	g
36	1	"	"	"	"	"	"
37	1	"	5	15	g	f	h
38	1	"	"	"	"	h	f
39	2	"	15	5	"	f	h
40	1	"	"	"	"	"	"
41	2	"	"	"	"	h	f
42	1	"	"	"	"	"	"
43	1	"	5	15	h	f	g
44	1	"	"	"	"	g	f
45	2	"	15	5	"	f	g
46	1	"	"	"	"	"	"
47	2	"	"	"	"	g	f
48	1	"	"	"	"	"	"

TABLE 6-R

Note: Same as Table 6-P, except enter this table with the value r.

Code Number r	Basic Square to be used	C	D	E	c	d	e
1	2	5	20	10	f	g	h
2	2	"	"	"	"	h	g
3	2	"	"	"	g	f	h
4	2	"	"	"	"	h	f
5	2	"	"	"	h	f	g
6	2	"	"	"	"	g	f
7	2	10	20	5	f	g	h
8	2	"	"	"	"	h	g
9	2	"	"	"	g	f	h
10	2	"	"	"	"	h	f
11	2	"	"	"	h	f	g
12	2	"	"	"	"	g	f
13	1	20	5	10	f	g	h
14	1	"	"	"	"	h	g
15	1	"	10	5	"	g	h
16	1	"	"	"	"	h	g
17	1	"	5	10	g	f	h
18	1	"	"	"	"	h	f
19	1	"	10	5	"	f	h
20	1	"	"	"	"	h	f
21	1	"	5	10	h	f	g
22	1	"	"	"	"	g	f
23	1	"	10	5	"	f	g
24	1	"	"	"	"	g	f

TABLE 6-S
Note: Same as Table 6-P, except enter this table
with the value *s*.

Code Number *s*	Basic Square to be used	*C*	*D*	*E*	*c*	*d*	*e*
1	1	10	15	20	*f*	*g*	*h*
2	1	"	"	"	"	*h*	*g*
3	1	"	20	15	"	*g*	*h*
4	1	"	"	"	"	*h*	*g*
5	1	"	15	20	*g*	*f*	*h*
6	1	"	"	"	"	*h*	*f*
7	1	"	20	15	"	*f*	*h*
8	1	"	"	"	"	*h*	*f*
9	1	"	15	20	*h*	*f*	*g*
10	1	"	"	"	"	*g*	*f*
11	1	"	20	15	"	*f*	*g*
12	1	"	"	"	"	*g*	*f*
13	1	15	10	20	*f*	*g*	*h*
14	1	"	"	"	"	*h*	*g*
15	1	"	20	10	"	*g*	*h*
16	1	"	"	"	"	*h*	*g*
17	1	"	10	20	*g*	*f*	*h*
18	1	"	"	"	"	*h*	*f*
19	1	"	20	10	"	*f*	*h*
20	1	"	"	"	"	*h*	*f*
21	1	"	10	20	*h*	*f*	*g*
22	1	"	"	"	"	*g*	*f*
23	1	"	20	10	"	*f*	*g*
24	1	"	"	"	"	*g*	*f*
25	1	20	10	15	*f*	*g*	*h*
26	1	"	"	"	"	*h*	*g*
27	1	"	15	10	"	*g*	*h*
28	1	"	"	"	"	*h*	*g*
29	1	"	10	15	*g*	*f*	*h*
30	1	"	"	"	"	*h*	*f*
31	1	"	15	10	"	*f*	*h*
32	1	"	"	"	"	*h*	*f*
33	1	"	10	15	*h*	*f*	*g*
34	1	"	"	"	"	*g*	*f*
35	1	"	15	10	"	*f*	*g*
36	1	"	"	"	"	*g*	*f*

TABLE 7
Fifth-Order Essentially Different Pandiagonal Magic
Squares

Consecutive No.	Frénicle No.	Basic Square to be used	A	B	C	D	E	a	b	c	d	e
1	1	1	0	5	10	15	20	1	2	3	4	5
2	3	1			"	15	20			"	5	4
3	5	1			"	20	15			"	4	5
4	7	1			"	20	15			"	5	4
5	9	1			"	15	20			4	3	5
6	11	1			"	15	20			"	5	3
7	13	1			"	20	15			"	3	5
8	15	1			"	20	15			"	5	3
9	17	1			"	15	20			5	3	4
10	19	1			"	15	20			"	4	3
11	21	1			"	20	15			"	3	4
12	23	1			"	20	15			"	4	3
13	25	2		15	10	20				3	4	5
14	27	2		"	10	20				"	5	4
15	29	1		"	20	10				"	4	5
16	31	1		"	20	10				"	5	4
17	33	2		"	10	20				4	3	5
18	35	2		"	10	20				"	5	3
19	37	1		"	20	10				"	3	5
20	39	1		"	20	10				"	5	3
21	41	2		"	10	20				5	3	4
22	43	2		"	10	20				"	4	3
23	45	1		"	20	10				"	3	4
24	47	1		"	20	10				"	4	3
25	49	2		20	10	15				3	4	5
26	51	2		"	10	15				"	5	4
27	53	2		"	15	10				"	4	5
28	55	2		"	15	10				"	5	4
29	57	2		"	10	15				4	3	5
30	59	2		"	10	15				"	5	3
31	61	2		"	15	10				"	3	5
32	63	2		"	15	10				"	5	3
33	65	2		"	10	15				5	3	4
34	67	2		"	10	15				"	4	3
35	69	2		"	15	10				"	3	4
36	71	2		"	15	10				"	4	3

PANDIAGONAL MAGIC SQUARES OF THE nth ORDER— n A PRIME NUMBER

A cyclical pandiagonal magic square is one, as we saw earlier, in which the capital and lower-case letters in each row, column and diagonal are cyclically arranged.

The number of cyclical pandiagonal magic squares of the nth order (where n is an odd prime) is given by the following formula (this formula is well known in the literature, but we have not been able to determine who proved it initially):

$$(n-3)(n-4)(n!)^2/8.$$

While it is entirely possible to prove this formula by straightforward mathematical methods, the following proof, based upon our new cyclical method, is much simpler:

Assume the number $(A+a)$ is in the upper left-hand cell. Then for the end result to be a cyclical pandiagonal magic square it is necessary that A appear somewhere in the second column from the left. Just what cell it occupies is a direct function of C and R. The question is, how many such combinations are there? Clearly, any two pairs of values which result in A occupying the same cell in the second column are equivalent from this point of view. Equally clearly, combinations which result in A falling in the top row, the next-to-top row or the bottom row must be eliminated as any of these combinations will prevent the final square from being pandiagonal. This leaves us with $(n-3)$ possibilities. A similar line of reasoning applies to the lower-case letter a except that here the values assigned to c and r determine the results we get. In addition, we must eliminate the cell occupied by the letter A, for if not we would have the number $(A+a)$ appearing more than once in the final square. Therefore, there are $(n-4)$ ways to position a for each of the $(n-3)$ ways of positioning A. This gives us a total of $(n-3)(n-4)$ possible combinations of the key numbers C, R, c and r. However, when we compare the resulting squares when $C=c$, $R=r$, $c=C$ and $r=R$, we find that the two are actually the same square—if you reverse the capital and lower-case letters you will get identical squares. In other words, there are exactly $(n-3)(n-4)/2$ distinctly different intermediate squares that will generate cyclical pandiagonal magic squares of the nth order—n a prime number. Pausing a moment to check back to the preceding chapter, we find that this is in agreement with the result we found there, since $(5-3)(5-4)/2=1$. We also saw in Chapter 9 that given one such fifth-order square we could generate from it $(5!)^2/4$—or in more general terms, $(n!)^2/4$—pandiagonal magic squares of order n. This gives us the required total of

$$(n-3)(n-4)(n!)^2/8$$

different cyclical pandiagonal magic squares of the nth order.

Considering next the seventh order, we see that there are $(7-3)$ $(7-4)/2=6$ distinct intermediate squares, each of which can give us $(7!)^2/4=6,350,400$ cyclical pandiagonal magic squares of the seventh order, for a grand total of 38,102,400 such squares in all. Interestingly enough, it is possible to construct tables, larger but similar in scope to those in the preceding chapter, which will enable you to construct at will the seventh-order pandiagonal magic square with any given Frénicle index number. Also it is entirely possible, though mechanically much more difficult than in the case of the fifth order, to list the Frénicle numbers of the 129,600 essentially different pandiagonal magic squares of the seventh order.

While, theoretically, higher-order pandiagonal magic squares could be treated in a similar fashion, for all practical purposes the task is an impossible one. There are 11,153,456,680,000 cyclical pandiagonal magic squares of the eleventh order alone!

Returning now to Professor Candy's irregular pandiagonal magic square of the seventh order, we stated in Chapter 14 that by the use of the two intermediate squares, Figures 59 and 60, you could construct all of the 640,120,320 magic squares of this type. It is a simple matter to show that this is so.

There are nine different ways six numbers can be selected from the set 1, 2, 3, 4, 5, 6 and 7, and still have $(p-q)=(r-s)=(t-u)$, namely,

$$2-1=4-3=6-5$$
$$2-1=4-3=7-6$$
$$2-1=5-4=7-6$$
$$3-2=5-4=7-6$$
$$3-1=4-2=7-5$$
$$3-1=6-4=7-5$$
$$4-1=5-2=6-3$$
$$5-2=6-3=7-4$$
$$5-1=6-2=7-3$$

For each of these nine combinations there are twelve ways they can be assigned to the lower-case letters a, b, c, d, e, f and g so that the relationship $(a-b)=(c-d)=(e-f)$ holds true.

There are, of course, seven factorial (or 5040) different ways the values 0, 7, 14, 21, 28, 35 and 42 can be assigned to the capital letters. This gives us $9 \times 12 \times 5040 = 544,320$ different squares so far per basic square. This number can be doubled by assigning the values 0, 7, 14, 21, 28, 35 and 42 to the lower-case letters (which can also be done in $9 \times 12 = 108$ different ways) and the values 1, 2, 3, 4, 5, 6 and 7 to the capital letters (which can, of course, be done in 5040 different ways). This total can again be doubled by reversing the rows and columns with the diagonals. This total can be tripled by permuting the rows and columns in the order 1-3-5-7-2-4-6 and in the order 1-4-7-3-6-2-5. This gives us 6,531,840 different squares per basic square so far.

But remember there are $n^2 = 49$ different cyclic variations of each different square. This gives us $6,531,840 \times 49 = 320,060,160$ different irregular pandiagonal magic squares per basic square. We can, therefore, construct 640,120,320 such squares using Figures 59 and 60! If anyone discovers another basic square he will increase this number by 320,060,160!

MATHEMATICAL PROOFS OF THE VARIOUS METHODS

PART IV

MATHEMATICAL
PROOFS OF THE
VARIOUS METHODS

21

INTRODUCTORY

This section of the book actually needs no introduction. Its purpose is self-evident. It is felt that to include the proof of the sufficiency of the various methods in the same chapters with the discussion of the methods themselves would be very confusing to the reader who was interested primarily in the end result. On the other hand, it was also felt that not to include a discussion of the basic theory in the book would be unfair to the reader interested in understanding what makes magic squares tick and how really satisfying they can be. The question, "Given a set of desired properties, can you construct a square to meet them and, if so, how?" has already been discussed. The question now facing us is, "Why does the method work?"

In dealing with this problem it will, at times, be necessary to assume a certain familiarity on the part of the reader with the properties of arithmetical series and of congruencies. The amount of advanced mathematics employed will, however, be kept to an absolute minimum.

22

TRANSPOSITION

We have already seen that rearranging the order of the rows and/or columns has considerable effect upon the magical properties of a square. Clearly, as long as rows are interchanged only with other rows and columns only with other columns, the magical properties of the rows and columns are not affected. The situation is vastly different with the diagonals. Consider the case of a symmetrical magic square.

In Figure 82 columns A and B are symmetrically located about the center of the square and rows a and b are similarly located. For example, in an eighth-order magic square, if A was the second column from the left-hand side of the square, B would be the second column from the right-hand side, a would be the second row from the top and b would be the second row from the bottom. Hence e and d are symmetrically located on the upward main diagonal and complementary with the sum (n^2+1). Likewise c and f are symmetrically located on the downward main diagonal and are complementary with the sum (n^2+1).

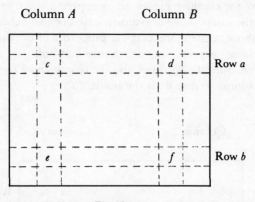

Fig. 82.

If we interchange only the columns we get Figure 83. Notice that while the main diagonals are different, because of the symmetrical property of the original square the totals have not changed and the main diagonals are still correct. Nor, for that matter, has the property of symmetry as a whole been disturbed.

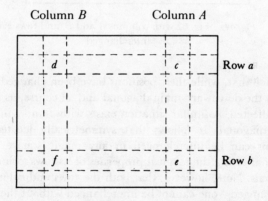

Fig. 83.   Fig. 82 with columns A and B interchanged.

In other words, symmetrically located rows and/or columns can be interchanged in a symmetrical magic square without adversely affecting the magical properties of the rows, columns or main diagonals, or its property of symmetry. It would, of course, affect any other magical properties that the square may have.

If next we consider Figure 82 as applying to an ordinary magic square, the interchange of symmetrically located columns would, in all likelihood, destroy the magical properties of the main diagonals. Contrariwise, suppose we interchange not only the symmetrically located columns *A* and *B* but also the rows *a* and *b* corresponding to those columns. Figure 84 is the result.

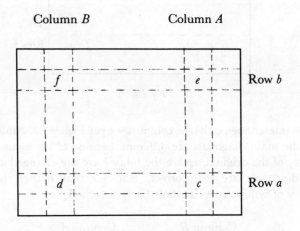

Fig. 84.   Fig. 82 with columns *A* and *B*, and rows *a* and *b* interchanged

Notice that, while their positions have been changed, *c* and *f* are still on the downward main diagonal and, of course, its total has not been affected. A similar situation exists with *d* and *e* in the upward main diagonal. It follows that symmetrically located rows and columns can be interchanged in any magic square without adversely affecting the magical properties of its rows, columns or main diagonals. Note, however, that both the rows and columns must be interchanged—one cannot be interchanged without the other unless the original square was a symmetrical one.

With pandiagonal magic squares the situation is still different. In Chapter 19 we said, "It follows, therefore, that we have the original square, the 1-3-5-2-4 transformation thereof, the interchange of the rows and columns with the diagonals for both of these squares, and the 24 additional squares we can get from each of these four squares by cyclical permutations, for a grand total of 100 pandiagonal magic squares that can be obtained from any pandiagonal magic square by means of transformations."

In order to prove the first part of this statement, let us begin with Figure 85a (which is identical with our basic fifth-order magic square, Figure 78). If we transpose the rows and columns in the order 1-3-5-2-4 we get Figure 85b. If we further interchange the rows and columns of both of these squares with their diagonals we generate Figures 85c and 85d.

$A+a$	$B+b$	$C+c$	$D+d$	$E+e$
$C+d$	$D+e$	$E+a$	$A+b$	$B+c$
$E+b$	$A+c$	$B+d$	$C+e$	$D+a$
$B+e$	$C+a$	$D+b$	$E+c$	$A+d$
$D+c$	$E+d$	$A+e$	$B+a$	$C+b$

Fig. 85a. The original square.

$A+a$	$C+c$	$E+e$	$B+b$	$D+d$
$E+b$	$B+d$	$D+a$	$A+c$	$C+e$
$D+c$	$A+e$	$C+b$	$E+d$	$B+a$
$C+d$	$E+a$	$B+c$	$D+e$	$A+b$
$B+e$	$D+b$	$A+d$	$C+a$	$E+c$

Fig. 85b. Fig. 85a with the rows and columns transposed in the order 1-3-5-2-4.

$A+a$	$B+c$	$C+e$	$D+b$	$E+d$
$C+b$	$D+d$	$E+a$	$A+c$	$B+e$
$E+c$	$A+e$	$B+b$	$C+d$	$D+a$
$B+d$	$C+a$	$D+c$	$E+e$	$A+b$
$D+e$	$E+b$	$A+d$	$B+a$	$C+c$

Fig. 85c. Fig. 85a with the rows and columns interchanged with the diagonals.

$A+a$	$B+d$	$C+b$	$D+e$	$E+c$
$C+e$	$D+c$	$E+a$	$A+d$	$B+b$
$E+d$	$A+b$	$B+e$	$C+c$	$D+a$
$B+c$	$C+a$	$D+d$	$E+b$	$A+e$
$D+b$	$E+e$	$A+c$	$B+a$	$C+d$

Fig. 85d. Fig. 85b with the rows and columns
interchanged with the diagonals.

Inspection will show that all four of these squares have each
capital letter and each lower-case letter appearing once, and only
once, in each row, column and diagonal. They are, therefore,
pandiagonal magic squares regardless of the values assigned to the
letters.

For reasons that will soon be evident let us permute the rows in
Figure 85b in the order 2-3-4-5-1 (that is, move the top row to the
bottom of the square by a cyclical permutation) to get Figure 85e.
Now rotate Figure 85e 90° clockwise to get the equivalent square,
Figure 85f.

$E+b$	$B+d$	$D+a$	$A+c$	$C+e$
$D+c$	$A+e$	$C+b$	$E+d$	$B+a$
$C+d$	$E+a$	$B+c$	$D+e$	$A+b$
$B+e$	$D+b$	$A+d$	$C+a$	$E+c$
$A+a$	$C+c$	$E+e$	$B+b$	$D+d$

Fig. 85e. Fig. 85b with the top row permuted to the
bottom of the square.

$A+a$	$B+e$	$C+d$	$D+c$	$E+b$
$C+c$	$D+b$	$E+a$	$A+e$	$B+d$
$E+e$	$A+d$	$B+c$	$C+b$	$D+a$
$B+b$	$C+a$	$D+e$	$E+d$	$A+c$
$D+d$	$E+c$	$A+b$	$B+a$	$C+e$

Fig. 85f. Fig. 85e rotated 90° clockwise.

Since Figure 85f was derived from Figure 85b by a cyclical permutation and a rotation, it follows that Figures 85f and 85b belong to the same set of 25 squares obtainable from a given square by cyclical permutation.

Now examine Figures 85a, 85f (which can be derived from Figure 85b), 85c and 85d. They all have the capital letters in identical positions and the lower-case letters in different positions. Obviously, no one of the four can be obtained from one of the others by cyclically permuting the rows and/or columns.

It remains only to show that cyclically permuting the rows and/or columns does not disturb the pandiagonal property of the square. An examination of Figure 85e, which (as we just saw) was derived from Figure 85b by permuting the rows in the cyclical order 2-3-4-5-1, will show that here again each capital and lower-case letter appears once, and only once, in each row, column and diagonal and that the square is a pandiagonal magic square, as required. If we can do this once we can repeat it as often as we wish and similarly with the columns. Thus we can derive 100 different fifth-order pandiagonal magic squares from any given fifth-order pandiagonal magic square.

As we saw in Chapter 20, in addition to transposing the rows and columns with the diagonals there are two noncyclical transpositions (1-3-5-7-2-4-6 and 1-4-7-3-6-2-5) which will leave any seventh-order pandiagonal magic square with all of its rows, columns and diagonals correct.

While not too difficult, it is a bit involved to prove, and we will merely state it: For any order n (n a prime number) there are $(n-3)/2$ transpositions possible with this property. If we include the original square, this gives us $(n-1)/2$ squares. By multiplying this by 2 for the transposition of the rows and columns with the diagonals and by n^2 for the cyclical transpositions, we find that there are, for any n (n a prime number), $(n-1)(n^2)$ different pandiagonal magic squares that can be derived from any given nth-order pandiagonal magic square. For any particular n, this statement can be verified by utilizing the method used above in proving it for the fifth order.

23

DOUBLY-EVEN ORDER—
COMPLEMENTARY METHOD

The method described in Chapter 2 is actually a special case of a more general method. We will, therefore, treat the more general case first. Since there is little loss in generality, and it is much easier to do, we shall treat the case of an eighth-order square. The extension to any doubly-even-order square will be obvious.

Given any arithmetical progression with initial term a and difference d, form a construction square by writing the numbers in their normal order as shown in Figure 86a.

By the very nature of the construction of this square it is a symmetrical square, that is, the sum of any two numbers symmetrically located with respect to the center equals $[2a + d(n^2 - 1)]$. The sum of n^2 terms of an arithmerical progression is $[2a + d(n^2 - 1)]$ $n^2/2$. The magic constant is, of course, one nth of this, or $[2a + d(n^2 - 1)]n/2$. It follows that Figure 86a is magic so far as the main diagonals are concerned (this will be true regardless of the value of n).

$a+0$	$a+d$	$a+2d$	$a+3d$	$a+4d$	$a+5d$	$a+6d$	$a+7d$
$a+8d$	$a+9d$	$a+10d$	$a+11d$	$a+12d$	$a+13d$	$a+14d$	$a+15d$
$a+16d$	$a+17d$	$a+18d$	$a+19d$	$a+20d$	$a+21d$	$a+22d$	$a+23d$
$a+24d$	$a+25d$	$a+26d$	$a+27d$	$a+28d$	$a+29d$	$a+30d$	$a+31d$
$a+32d$	$a+33d$	$a+34d$	$a+35d$	$a+36d$	$a+37d$	$a+38d$	$a+39d$
$a+40d$	$a+41d$	$a+42d$	$a+43d$	$a+44d$	$a+45d$	$a+46d$	$a+47d$
$a+48d$	$a+49d$	$a+50d$	$a+51d$	$a+52d$	$a+53d$	$a+54d$	$a+55d$
$a+56d$	$a+57d$	$a+58d$	$a+59d$	$a+60d$	$a+61d$	$a+62d$	$a+63d$

Fig. 86a.

Consider any two symmetrically located columns. There is a constant difference between the terms in any given row. For example, in columns 0 and 7, the difference in any given row is $7d$. Thus the total of column 7 is n (8 in our example) times this difference greater than the total of column 0. If we transfer one-half of the numbers in column 7 with their corresponding numbers in column 0 (being careful to interchange only numbers in the same row) we will make columns 0 and 7 magic without affecting the totals of the rows. If we make similar interchanges between the numbers in columns 1 and 6, 2 and 5, and 3 and 4, we will have all columns magic. The same principle holds between rows. If we interchange one-half of the numbers in any row with those in its symmetrical row (again being careful to interchange only numbers in the same column), we will make all rows magic without affecting the totals of the columns. This gets quite complicated, however, particularly so as we must be careful not to change the totals of the main diagonals.

Fortunately there is a simplifying device available to us. Instead of following the above procedure, interchange numbers with their complements according to the rule: whenever you interchange a number, say one in column w and row x, with its complement which happens to be in column y and row z, you also interchange the number in column w and row z with its complement which you will find in column y and row x. In other words, you simultaneously make a partial correction to both columns and rows involved without affecting the totals of the main diagonals. All you have to do is to follow the above rule and interchange one-half the numbers in the square—selected so that $n/2$ of them are in each row and $n/2$ in each column. The resultant square will automatically be magic as well as symmetrical. Needless to say, there are many ways this can be done, the crossed diagonals of Chapter 2 being only one of them. However, since it is very simple and straightforward, it is frequently used.

If you set $a = 1$, $d = 1$, and interchange the numbers on the crossed diagonals, you will get Figure 11 of Chapter 2.

If you set $a = 1$, $d = 2$, and interchange the numbers set off in the blocks in Figure 86a, you will generate Figure 86b—a symmetrical eighth-order magic square formed of the first 64 odd numbers.

1	3	123	121	119	117	13	15
17	19	107	105	103	101	29	31
95	93	37	39	41	43	83	81
79	77	53	55	57	59	67	65
63	61	69	71	73	75	51	49
47	45	85	87	89	91	35	33
97	99	27	25	23	21	109	111
113	115	11	9	7	5	125	127

Fig. 86b.   A symmetrical eighth-order magic square formed of the first 64 odd numbers.

24

MODIFIED STRACHEY
METHOD

Even though the method for generating singly-even magic squares described in Chapter 4 was an original generalization of Strachey's method, we shall, as in the case of the method just described for doubly-even magic squares, prove a still more general method.

Strachey's method consists basically in superimposing a magic square of order $n/2$ in each of the four quadrants of a magic square of order n. The question naturally arises, "What are the minimum requirements that must be met by these two squares for the resultant square (formed by adding together the two numbers in each cell) to be magic?"

One of these requirements is that each of the two squares must be magic for, unless they are both magic, there is not much chance that the resultant square will be magic.

So far so good, but we are interested in normal magic squares formed of the numbers 1 to n^2, inclusive, each used once, and only once. Since the square of order $n/2$ will consist of the numbers 1 to

$n^2/4$ if it is a normal square, this becomes another of our requirements. The next requirement follows directly from it. The construction square of order n must be formed of the numbers 0, $n^2/4$, $n^2/2$ and $3n^2/4$ so arranged that, when the squares are superimposed, any one number in the smaller square will be combined once, and only once, with each of the four numbers forming the larger square. There is no question but that the rules given in Chapter 4 meet these requirements. The point is, they are unnecessarily restrictive.

Combine the normal magic square of order $n/2$ into a square of order n as shown in Figure 87. In constructing an even-order magic square there is no need to write this square out. We are doing it here to show the exact manner in which the $n/2$ order square is used.

a	b	c ···	k	a	b	c ···	k
·				·			
·	Square in			·	Square in		
·	normal			·	normal		
·	position			·	position		
p	q	r ···	w	p	q	r ···	w
p	q	r ···	w	p	q	r ···	w
·				·			
·	Square mir-			·	Square mir-		
·	rored about			·	rored about		
·	horizontal axis			·	horizontal axis		
a	b	c ···	k	a	b	c ···	k

Fig. 87.

Clearly, if the original square of order $n/2$ was magic, then Figure 87 will be magic. If the original square was pandiagonal, then Figure 87 will be pandiagonal.

Remembering that by construction the large square is also symmetrical about the same horizontal axis, it is readily seen that any given number, say a, will be combined with 0, $n^2/4$, $n^2/2$ and $3n^2/4$, as required, provided only that any pair of numbers (in the large construction square) in the same row which are $n/2$ columns apart are not identical or complementary. That is, if 0 or $3n^2/4$ is in cell $[i,j]$ then the number in cell $[i+n/2,j]$ must be either $n^2/2$ or $n^2/4$.

The columns will be magic automatically because of the symmetrical construction of the square. For the rows to be magic it is necessary for the sum of the quantities in each row to equal $(3n/2)(n^2/4)$. This is easily done by placing each number in a row as indicated in the following table:

	Place the number heading the column in any given row the number of times			
When	indicated below it:			
$n=2(2m+1)$	0	$n^2/4$	$n^2/2$	$3n^2/4$
Use either this	m	$m+2$	$m-1$	$m+1$
or this combination	$m+1$	$m-1$	$m+2$	m
$n=2(2m)$ use	m	m	m	m

For the main diagonals to be magic all that is necessary is for the portions of the main diagonals in the upper quadrants to be equal. The fact that the square is symmetrical about the horizontal axis will take care of the rest!

To summarize: The minimum requirements are:

REQUIREMENT A: Form an nth-order magic square by combining four identical, normal magic squares of $n/2$ order as shown in Figure 87. (As mentioned before, this square need not be written down as a separate entity. All that is necessary is that, when it is superimposed over the main construction square, the $n/2$ square be placed as indicated in Figure 87.)

REQUIREMENT B: Form an nth-order magic square using the numbers 0, $n^2/4$, $n^2/2$ and $3n^2/4$ so arranged as to meet the following conditions:

(1) The lower half is symmetrical to the upper half, that is, it is a mirror of the upper half with every number being replaced by its complement—0 by $3n^2/4$, $n^2/4$ by $n^2/2$, $n^2/2$ by $n^2/4$ and $3n^2/4$ by 0.

(2) In each row the numbers which are $n/2$ columns apart are not identical or complementary.

(3) The sum of the terms in the rows in the upper half is $(3n/2)(n^2/4)$. (Following the above table will ensure that this requirement is met.)

(4) The totals of the portions of the main diagonals lying above the horizontal dividing line are equal to each other.

Let us demonstrate the wide flexibility of the above method by constructing a sixth-order magic square (which is usually a quite restricted process) by the aid of Figure 1. This gives us Figure 88a for our first construction square. It is clearly magic.

8	1	6	8	1	6
3	5	7	3	5	7
4	9	2	4	9	2
4	9	2	4	9	2
3	5	7	3	5	7
8	1	6	8	1	6

Fig. 88a.

Since $n^2/4=9$, the four numbers we shall use for the second requirement are 0, 9, 18 and 27. There are many different ways we can arrange these four numbers so as to meet the four conditions listed under Requirement B. Figure 88b is merely one of them.

9	9	9	27	27	0
27	9	27	9	0	9
0	18	0	18	27	18
27	9	27	9	0	9
0	18	0	18	27	18
18	18	18	0	0	27

Fig. 88b.

Superimposing Figure 88a on Figure 88b and adding together the two numbers in each cell gives us the final square, Figure 88c, a sixth-order magic square.

17	10	15	35	28	6
30	14	34	12	5	16
4	27	2	22	36	20
31	18	29	13	9	11
3	23	7	21	32	25
26	19	24	8	1	33

Fig. 88c.   A sixth-order magic square.

As mentioned earlier, pandiagonal magic squares of the singly-even order cannot exist. The following proof, adapted from that given by C. Planck,* is quite simple. Take any even square of the order $2m$ and make a double checkerboard of it as shown in Figure 89, where there are four—rather than the normal two—different types of squares systematically arranged.

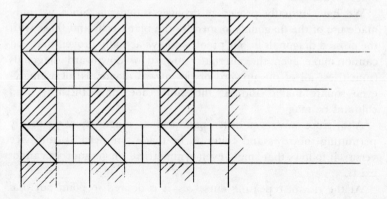

Fig. 89.

---

*"Pandiagonal Magics of Order 6 and 10 with Minimal Numbers" by C. Planck, *Monist* XXIX, 307–316, 1919.

Assume that this square is filled with the first $(2m)^2$ natural numbers so arranged that the square is a normal pandiagonal magic square. The magic constant then becomes $m(4m^2+1)=S$. Let the sum of the numbers in the shaded cells be A, in the cells with the diagonal lines B, and in the blank cells C. Then we have the sum of the numbers in the 1st, 3rd, 5th etc. rows equal to $A+C=mS$. Similarly, we have the sum of the numbers in the 1st, 3rd, 5th etc. columns equal to $A+B=mS$. Likewise, the sum of the numbers in the alternate diagonals formed by the blank cells and the cells with the diagonal lines equals $C+B=mS$.

These equations give us:

$$A+B=A+C=B+C=mS.$$

This equation can be true only if:

$$mS/2=A=B=C=m^2(4m^2+1)/2=2m^4+m^2/2.$$

But, when the square is singly even, m is odd and we will have: $A=B=C=$ a fraction. Since this is not possible—remember, all of the cells are occupied by whole numbers—it follows that our original assumption—that the natural numbers 1 to $(2m)^2$, inclusive, were arranged so as to form a normal pandiagonal magic square—is false when the square is singly even. That is to say, pandiagonal magic squares of the singly-even order cannot exist!

We have actually proved a stronger statement. Since we only made use of the diagonals formed by the blank cells and those with the crossed diagonals in our proof, we have actually proven that you cannot make even the alternate upward or downward diagonals (much less all of the upward or downward diagonals) of a singly-even square magic and, at the same time, have the rows and columns be magic.

Also, since any even-order pandiagonal magic square can, by permuting the rows and columns, be made symmetrical (and vice versa), it follows that singly-even symmetrical magic squares cannot exist!

At the risk of repeating ourselves, it is desired to point out that this Modified Strachey Method (as described herein) may be used to construct doubly-even, as well as singly-even, magic squares of any order above four.

25

BORDERED MAGIC SQUARES

While the proof of the method used to generate bordered magic squares in Chapter 5 is not difficult, it is tedious. For this reason, and the fact that the examples previously shown demonstrate that it does produce normal magic squares, the proof will not be given in its entirety. However, it will be sufficiently indicated so that anyone interested can complete the proof for himself.

To show that the square will be magic is quite simple. Remembering that the magic constant for any normal square of order n is $n(n^2+1)/2$, we see that the magic constant for the nucleus magic square of order $(n-2)$ is $(n-2)[(n-2)^2+1]/2$. After we add $2(n-1)$ to each term of this square the magic constant becomes:

$$(n-2)\left[(n-2)^2+1\right]/2 + 2(n-1)(n-2) = (n-2)(n^2+1)/2.$$

Since the method calls for all numbers, other than the corner numbers, that lie in the same row or column to be complementary

—and thus have a sum of (n^2+1)—it is seen that the total of the numbers in any row or column, other than the border rows and columns, will equal

$$(n-2)(n^2+1)/2+(n^2+1)=n(n^2+1)/2,$$

as required. Since diagonally opposite corner numbers are also complementary, a similar statement holds for the main diagonals.

In the case of the border rows or columns, all we have to do is to total the numbers listed. For example, in the case of Figure 26 we have the following total for the left-hand column:

$$n^2-(n-1)/2+n+(n-1)/2 \text{ plus total of Set A}$$

$$=n^2+n+n+n^2(n-3)/2+n(n-3)/2$$

$$=n^3/2+n/2=n(n^2+1)/2, \text{ as required.}$$

Similar checks will show that in all cases the border rows and columns will add up to the magic constant.

Also, investigation will show that all the numbers 1 to n^2 are used once, and only once, and that the square is a normal magic square.

26

NEW CYCLICAL METHOD

In the following proof we will use the numbers 1 to n^2, inclusive, rather than resorting to the use of the intermediate construction square. This will cause no loss in generality of the proof as it is equivalent to putting the numbers 1 to n^2 to the base n and then setting $A, B, C, \ldots, N$ equal to $0, n, 2n, \ldots, (n-1)n$, and $a, b, c, \ldots, n$ equal to $1, 2, 3, \ldots, n$. Clearly, this is a one-to-one transformation and can be made in either direction.

If we represent any number, say t, to the base n (where n is the order of the square), we have

$$t = Xn + x + 1,$$

where X and x are either positive integers less than n or are zero. Let (C, R) be the step used in constructing the square as you proceed from any number (other than those numbers which are an exact multiple of n) to the next consecutive number, let $(C + c, R + r)$ be the cross-step as you proceed from any number which is an exact multiple of n to the next consecutive number, and (i,j) the cell selected for the number 1. Then the step from 1 to t (that is, from 1 to $Xn + x + 1$) is $(Cx + cX, Rx + rX)$ and the number t will fall in cell $(i + Cx + cX, j + Rx + rX)$.

Similarly, any other number, say $t + Yn + y$, will fall in cell $(i + Cx + cX + Cy + cY, j + Rx + rX + Ry + rY)$. For these two numbers to lie in the same cell it follows that both $(Cy + cY)$ and $(Ry + rY)$ must equal zero modulo n. Hence if every number from 1 to n^2 is to occupy a different cell, both Y and y must equal zero whenever both

$$Cy + cY \equiv 0 \, (\text{mod} \, n) \quad \text{and} \quad Ry + rY \equiv 0 \, (\text{mod} \, n)$$

are true congruences. These two congruences may be written as

$$Cy \equiv - cY \, (\text{mod} \, n) \quad \text{and} \quad rY \equiv - Ry \, (\text{mod} \, n).$$

Since any two congruences may be multiplied together member by member (provided they are of the same modulus) the following congruence will be true whenever the two original congruences are true:

$$CyrY \equiv RycY \, (\text{mod} \, n) \quad \text{or} \quad (Cr - cR)yY \equiv 0 \, (\text{mod} \, n).$$

If $(Cr - cR)$ is prime to n the last congruence can be true only if y or Y equals zero. If we assume that Y equals zero, substituting this value in the original congruences gives:

$$Cy + 0 \equiv 0 \, (\text{mod} \, n) \quad \text{and} \quad Ry + 0 \equiv 0 \, (\text{mod} \, n).$$

If these congruences are true, either y equals zero or both C and R equal zero. If the latter is true, then

$$(Cr - cR) \equiv 0 \, (\text{mod} \, n),$$

which is contrary to the assumption that $(Cr - cR)$ is prime to n. It follows that if Y equals zero then y will also equal zero when the two original congruences are true and $(Cr - cR)$ is prime to n. A similar result will be obtained if it is originally assumed that y equals zero. We see, therefore, that if $(Cr - cR)$ is prime to n and if both

$$(Cy + cY) \equiv 0 \, (\text{mod} \, n) \quad \text{and} \quad (Ry + rY) \equiv 0 \, (\text{mod} \, n)$$

are true congruences, both y and Y will equal zero. Conversely, it is clear that if $(Cr - cR)$ is not prime to n, then values of y and Y exist for which two or more numbers will occupy the same cell. We see, therefore, that if we wish each of the numbers to occupy a separate cell, that $(Cr - cR)$ must be prime to n.

As noted above, the number t (where $t = Xn + x + 1$) will fall in the column $(i + Cx + cX)$. The n numbers of any given column, say column $(i + d)$, will be the n solutions of the congruence

$$Cx + cX \equiv d \pmod{n}.$$

From the theory of congruences it can be shown that when C and c are prime to n there will be n different combinations of

$$x = 0, 1, 2, 3, \ldots, (n - 1)$$

and

$$X = 0, 1, 2, 3, \ldots, (n - 1)$$

that will be solutions of this congruence. For example, let $C = 1$, $c = 5$, $d = 1$ and $n = 6$. Substituting the following values of X and x will show that they are solutions of the resultant congruence, $x + 5X \equiv 1 \pmod{6}$: $X = 0$, $x = 1$; $X = 1$, $x = 2$; $X = 2$, $x = 3$; $X = 3$, $x = 4$; $X = 4$, $x = 5$; and $X = 5$, $x = 0$.

If either C or c is not prime to n the situation is different. Assume C and n have a g.c.d (greatest common divisor) equal to s (where s is not equal to 1). Now there will be no solution to the above congruence unless $(d - cX)$ is also divisible by s. In this case there will be s different values of x corresponding to the value of X. For example, let $C = 3$, $c = 5$, $d = 1$ and $n = 6$. Substituting the following values of X and x will show that they are solutions of the resultant congruence, $3x + 5X \equiv 1 \pmod{6}$: $X = 0$, $x =$ none; $X = 1$, $x =$ none; $X = 2$, $x = 1$; $X = 2$, $x = 3$; $X = 2$, $x = 5$; $X = 3$, $x =$ none; $X = 4$, $x =$ none; $X = 5$, $x = 0$; $X = 5$, $x = 2$; $X = 5$, $x = 4$.

The same principle holds when both C and c are not prime to n. Let $C = 3$, $c = 2$, $d = 1$ and $n = 6$, giving us $3x + 2X \equiv 1 \pmod{6}$. The solutions are: $X = 2$, $x = 1$; $X = 2$, $x = 3$; $X = 2$, $x = 5$; $X = 5$, $x = 1$; $X = 5$, $x = 3$; and $X = 5$, $x = 5$.

A similar analysis can be made of the rows, the n numbers in any given row, say row $(j + e)$, being the n solutions of the congruence:

$$Rx + rX \equiv e \pmod{n}.$$

It is thus seen that the number of different capital letters (or lower-case letters) in any given row (or column) will depend upon the g.c.d. of the controlling characteristic (R, r, C or c, as the case may be) and n. For example, in the case of the capital letters in the columns, it would be the g.c.d. of C and n. If the g.c.d. is 1 (that is, if the two numbers are prime to each other) all the n capital letters in the columns will be different. If the g.c.d. is 2, there will be $n/2$ different capital letters, each appearing two times. If the g.c.d. is 3, there will be $n/3$ different capital letters, each appearing three times. And so forth up to the case where the g.c.d. is n (that is, where C is equal to zero). In this case the same capital letter will appear in all of the cells of any given column.

The sufficiency of Condition 1 (as stated on page 61) is thus apparent. It is also apparent that C is the controlling characteristic for the capital letters in the columns, c for the lower-case letters in the columns, R for the capital letters in the rows and r for the lower-case letters in the rows.

It remains to show that $(qC - pR)$ is the controlling characteristic for the capital letters in the $\{p,q\}$ series and $(qc - pr)$ for the lower-case letters in the same series.

Assume a specific $\{p,q\}$ series is started from any cell, say (g,h), selected at random. Then the n numbers involved in that particular series will occupy the n cells $(g + kp, h + kq)$, where k takes on the n values $0, 1, 2, 3, \ldots, (n-1)$. If p times the row number of any cell in the series is subtracted from q times the column number of the same cell the result is:

$$q(g + kp) - p(h + kq) = qg + qkp - ph - pkq = qg - ph$$

and, since this result is independent of the cell selected, it must be a constant, hence

$$qg - ph \equiv f \pmod{n},$$

where f is some positive integer from 1 to $(n-1)$, inclusive, or zero.

If the number 1 is in the cell (i,j) and the number occupying cell (g,h) is $Zn + z + 1$, the following relations, as shown previously, hold:

$$g \equiv i + Cz + cZ \pmod{n} \quad \text{and} \quad h \equiv j + Rz + rZ \pmod{n}.$$

Substituting these values for g and h in the above congruence gives:

$$qg - ph \equiv q(i + Cz + cZ) - p(j + Rz + rZ) \equiv f \pmod{n}$$

or

$$(qC - pR)z + (qc - pr)Z \equiv f' \pmod{n},$$

where f' is some positive integer from 1 to $(n-1)$, inclusive, or zero.

Since, regardless of which cell (of the n cells forming any specific $\{p,q\}$ series) is used as a starting point, the same cells (and thus the same numbers) are contained in the series, it follows that the n solutions of the congruence

$$(qC - pR)x + (qc - pr)X \equiv f' \pmod{n}$$

will determine the n numbers in the specific $\{p,q\}$ series which includes the number $(Zn + z + 1)$. Clearly, the solutions of this congruence are determined in exactly the same manner as was the case with the congruence

$$Cx + cX \equiv d \pmod{n}.$$

The only difference is that here the controlling characteristic for X (that is, for the capital letters) is $(qC - pR)$, and for x (that is, for the lower-case letters) it is $(qc - pr)$. The controlling characteristics for the $\{1,1\}$ and the $\{1,-1\}$ series are readily determined by letting p and q equal these values. This gives us $(C - R)$ and $(c - r)$ for the upward diagonals and $(-C - R)$ and $(-c - r)$ for the downward diagonals. Note that, since the sign of the controlling characteristic is of no importance (what counts is the g.c.d. of the controlling characteristic and n) we write the latter as $(C + R)$ and $(c + r)$ as a matter of convenience.

Let us now investigate the conditions under which, when our square has a column, row, diagonal or $\{p,q\}$ series consisting of n/s different capital (or lower-case) letters, each appearing s times (where s is the g.c.d. of the controlling characteristic and n), the square can be made magic. Another way of stating the problem is, let us investigate the conditions under which we can divide the numbers 1 to n—or, what is the same thing, 0 to $(n-1)$—inclusive, into s sets of n/s each, with all sets having the same total.

The average value of the numbers 1 to n, inclusive, is $1/n$th their total, or $(n+1)/2$. While this number is an integer when n is odd, it is a fraction when n is even. It follows, therefore, that when n is even and n/s is odd it is not possible to arrange the numbers 1 to n, inclusive, into s sets (of n/s numbers each) with all sets having the same total.

What about when n and n/s are both even? This is quite easy. Form $n/2$ pairs of numbers as follows:

$$
\begin{array}{cccccc}
1 & 2 & 3 & 4 & \cdots & n/2 \\
n & n-1 & n-2 & n-3 & \cdots & n/2+1.
\end{array}
$$

Notice that the sum of the numbers forming each pair is $(n+1)$, or twice the average value of a single term. It follows that by selecting $n/2s$ pairs of these numbers we will have a set of n/s numbers totaling $1/s$ of the total of 1 to n, inclusive. The picking of s such sets may be done in many different ways.

We now return for a moment to singly-even squares. If C, R, c and r, are all odd $(Cr - cR)$ will be even and cannot be prime to n. In other words, you cannot construct a singly-even square without having at least one of the controlling characteristics even. It follows that for this characteristic, n/s will be odd and that numerical values cannot be selected from the numbers 1 to n—or 0 to $(n-1)$ as the case may be—inclusive, that will make the square magic. Hence, you must resort to the expedient of transferring some of the numbers, as shown in Chapter 9.

On the other hand, if the square is doubly even you will always be able to make the square magic if you are careful to select C, R, c and r such that n/s is even.

In the case where n is odd all factors must be odd and our problem reduces to, "When a and b are both odd and $a \times b = n$, can the numbers 1 to n, inclusive, be divided into a sets of b each, with each set having the same sum?" The answer is yes!

With the exception of the middle term k, where $k = (n+1)/2$, the numbers 1 to n, inclusive, can be written as $(k-1)$ pairs of numbers, each pair adding to $2k = (n+1)$, as follows:

$$
\begin{array}{cccccccc}
1 & 2 & 3 & \cdots & (k-i) & \cdots & (k-2) & (k-1)_k \\
(2k-1) & (2k-2) & (2k-3) & \cdots & (k+i) & \cdots & (k+2) & (k+1)
\end{array}
$$

where i is any integer from 1 to $(k-1)$.

Let $j = (a-1)/2$ and select $3a$ numbers as follows:

(1) the middle a numbers, namely, $(k-j)$ to $(k+j)$, inclusive,

(2) any other a consecutive numbers, say $(k-i)$ to $(k-i+2j)$, inclusive, and

(3) the a consecutive numbers which are complementary to those just selected, namely, $(k+i-2j)$ to $(k+i)$, inclusive.

This will leave $(n-3a)/2$ pairs of complementary numbers available for future use.

Draw a blank rectangle having a columns and b rows. Insert the above $3a$ numbers in the first three rows as follows:

$$(k-i) \quad (k-i+1) \quad \cdots \quad (k-i+j) \quad (k-i+j+1) \quad \cdots \quad (k-i+2j)$$
$$k \qquad (k+1) \quad \cdots \quad (k+j) \qquad (k-j) \qquad \cdots \qquad (k-1)$$
$$(k+i) \quad (k+i-2) \quad \cdots \quad (k+i-2j) \quad (k+i-1) \quad \cdots \quad (k+i-2j+1)$$

A check will show that the three numbers in any column will add to $3k=3(n+1)/2$. If we fill in the remaining $(b-3)$ blank cells in each column with $(b-3)/2$ of the complementary pairs of numbers at our disposal, which we can do in many different ways, each column will consist of b numbers whose sum is $b(n+1)/2$, as required.

The above process looks quite formidable, but actually it is quite simple. Figure 90 shows one of the many different ways in which we can partition the numbers 1 to 35, inclusive, into seven sets of five numbers each with the total of the numbers in each set being 80. Notice that the total of the first three numbers of each column is 54 and that the total of the last two numbers of each column is 36.

3	4	5	6	7	8	9
18	19	20	21	15	16	17
33	31	29	27	32	30	28
25	1	10	23	2	14	24
11	35	26	13	34	22	12

Fig. 90.

Needless to say, it would have been just as simple to get five sets of seven numbers each with a total of 112 per set.

In view of the above, it is apparent that—as long as $(Cr-cR)$ is prime to n and no one of the controlling characteristics C, R, c and r

is equal to zero—any intermediate square of the odd order generated by our new cyclical method can be made magic by the proper selection of values for the letters.

The sufficiency of Condition 2 and the first part of Condition 3 on pages 61 and 62 is thus apparent.

It is this ability to assign values to the letters in such a manner as to make a row, column, diagonal or series magic that gives the method its great flexibility.

Now to return again to the singly-even square. We just noted that there was no way to avoid having n/s odd for at least one controlling characteristic. Let $C = 2$ and $n = 10$. Then there will be five capital letters in each column, each appearing two times. As an example, see Figure 41. It is necessary to interchange some of the individual numbers to make the sum of the capital letters in the columns correct. The question is, "What conditions must be met to make it possible to interchange two numbers in the same row so as to correct the totals of the capital letters in the columns involved without disturbing the total of the lower-case letters in the same two columns?"

Returning to Figure 41, we note that in the case of each interchange the numbers involved had the same lower-case letter. It is obvious that if you interchange $[B + b]$ with $[G + b]$ you will change the totals of the capital letters in the columns concerned but that no change will be made in the totals of the lower-case letters. In other words, in order to correct the totals of the capital letters in the columns (which consist of five different capital letters, each repeated two times) it is essential that the lower-case letters in the rows also consist of five different letters, each repeated two times. Hence the need for the last part of Condition 3 on page 62.

Turning now to Condition 4 (see page 64) we see that if $r = \pm C$ and $c = \pm R$ and y equals n divided by the g.c.d. of r and n (or, what is the same thing, the g.c.d. of C and n), the capital letters in the columns and the lower-case letters in the rows will both consist of y different letters each appearing n/y times. Examination of Figures 46 and 48 will show that not only is this true but also the letters always appear in the same cyclical order and if they are assigned values which will make the rows and columns correct, they will automatically make the y-by-y subsquares magic. A similar statement holds for the z-by-z subsquares where z equals n divided by the g.c.d. of c (or R) and n.

The proof of Condition 5 (see page 65) is not difficult but it is somewhat involved. Suffice it to say that the cyclical properties built into the square (by Condition 5) automatically take care of most of the requirements and that careful selection of the values for the letters will complete the task as shown by Figures 48 and 49.

27

DOUBLY-EVEN-ORDER
MAGIC SQUARES

The method employed in constructing doubly-even-order pandi-agonal magic squares in Chapter 11 is actually much easier to use than it is to describe or to prove.

Assume the number $[+1, -3]$ appears in one of the cells of the primary square. It will then appear in the upper left-hand corner of the main square, $[-1, -3]$ will appear in the same cell in the upper right-hand quadrant, $[+1, +3]$ in the same cell in the lower left-hand quadrant, and $[-1, +3]$ in the same cell in the lower right-hand quadrant. If Requirement A is not met it is clear that there will be duplication—the same number will appear in more than one cell. On the other hand, if Requirement A is met by the primary square every possible combination of $0, 1, 2, \ldots,$ $(n/2-1), -(n/2-1), \ldots, -2, -1, -0$ in the "tens" place with $0, 1, 2, \ldots, (n/2-1), -(n/2-1), \ldots, -2, -1, -0$ in the "units" place will appear once, and only once, in the main square when Step 1 is followed. It follows that the resultant square (after Steps 2,

3 and 4—the need for which is obvious) will consist of the numbers 1 to n^2, inclusive, as desired.

Let c, d represent any number, say q, to the base n, that is,

$$q = cn + d$$

and, as before, let $\pm v, \pm w$ represent any number in the "modified" nomenclature. That is, if v is positive then $c = v$ and if v is negative then $c = (n-1) - v$. Similarly, if w is positive then $d = w$ and if w is negative then $d = (n-1) - w$.

Also, we shall use the summation sign, Σ, in the same sense that it was used in Chapter 13.

By definition, for a square consisting of the numbers 0 to $(n^2 - 1)$, inclusive, to be magic it is necessary for the sum of the numbers in each rcd (row, column and main diagonal) to equal one-nth of the sum of the numbers 0 to $(n^2 - 1)$, that is, the following equation must be true for each rcd:

$$\Sigma(q) = n^2(n^2 - 1)/2 \div n = n(n^2 - 1)/2.$$

Obviously,

$$\Sigma(c) = \Sigma +, +(v) + \Sigma +, -(v) + \Sigma -, +(n-1-v)$$
$$+ \Sigma -, -(n-1-v)$$
$$= \Sigma(v) + r(n-1), \quad \text{where } r \text{ is the number of negative } v\text{'s.}$$

Similarly,

$$\Sigma(d) = \Sigma(w) + s(n-1), \quad \text{where } s \text{ is the number of negative } w\text{'s.}$$

If:

$$\Sigma(v) = \Sigma(w) = 0 \quad \text{and if}$$
$$r = s = (n)/2, \quad \text{then}$$
$$\Sigma(c) = \Sigma(d) = n(n-1)/2 \quad \text{and}$$
$$\Sigma(q) = \Sigma(cn + d) = n\Sigma(c) + \Sigma(d)$$
$$= n^2(n-1)/2 + n(n-1)/2 = (n^3 - n^2 + n^2 - n)/2$$
$$= n(n^2 - 1)/2, \quad \text{as required.}$$

In other words, if the conditions listed below are met by each rcd the resultant square will be magic (and, if met by the broken diagonals as well, it will be pandiagonal):

CONDITION 1:   The sum of all v's equals zero.

CONDITION 2:   The sum of all w's equals zero.

CONDITION 3:   There is an equal number of positive and negative v's.

CONDITION 4:   There is an equal number of positive and negative w's.

By the very nature of the generating square, Conditions 1 to 4, inclusive, will be met automatically in all cases except for the v's in the columns and the w's in the rows. If Requirements B to E, inclusive (see page 74), are met by the primary square, then Conditions 1 to 4, inclusive, will also be met by the v's in the columns and the w's in the rows and the square will be, as predicted, a doubly-even-order pandiagonal magic square.

28

BIMAGIC SQUARES

By definition, for a square to be bimagic it must be magic in both the first and second degrees. Hence it is necessary for the sum of the numbers in each rcd to equal $n(n^2-1)/2$ and for the sum of the squares of the numbers in each rcd to equal $n(n^2-1)(2n^2-1)/6$ when the numbers are expressed to the base 10. We are, of course, still talking about squares formed with the numbers 0 to (n^2-1), inclusive.

In the last chapter we saw that a square which was generated in accordance with Steps 1 to 4, inclusive, from the generating square shown in Figure 50 and a primary square which met Requirements A to E, inclusive, would meet Conditions 1 to 4, inclusive, and would be a first-degree pandiagonal magic square. It remains to establish the conditions which, if met, will result in the main square being magic in the second degree and to show that when Requirements F to J (see pages 78 and 79), inclusive, are met, these conditions will also be met. We shall use the same nomenclature that we used in the last chapter.

When the numbers are expressed to the base n we have:

$$\sum (q^2) = \sum (cn + d)^2 = n^2 \sum (c^2) + 2n \sum (cd) + \sum (d^2).$$

Substituting the "modified" nomenclature, we get:

$$\sum (c^2) = \sum {}^{+v}(v^2) + \sum {}^{-v}(n-1-v)^2$$
$$= \sum (v^2) + 2(n-1) \sum {}^{-v}(v) + r(n-1)^2.$$

Now assume that the following conditions are met and substitute these values in the equation for the sum of c^2.

CONDITION 3:   There is an equal number of positive and negative v's, that is, $r = n/2$.

CONDITION 5: $\sum (v^2) = n(n/2-1)(n-1)/6$.

CONDITION 6: $\sum -v(v) = -n(n/2-1)/4$.

This gives:

$$\sum (c^2) = n(n/2-1)(n-1)/6 - 2(n-1)(n)(n/2-1)/4$$
$$+ n(n-1)^2/2$$
$$= n(n-1)(n/2-1-3n/2+3+3n-3)/6$$
$$= n(n-1)(2n-1)/6.$$

In a similar manner it can be shown that the sum of d^2 will equal the same amount if the following conditions are met:

CONDITION 4:   There is an equal number of positive and negative w's, that is, $s = n/2$.

CONDITION 7: $\sum (w^2) = n(n/2-1)(n-1)/6$.

CONDITION 8: $\sum -w(w) = -n(n/2-1)/4$.

Considering next the cross-products:

$$\sum (cd) = \sum {}^{+,+}(vw) + \sum {}^{+,-}(v)(n-1-w)$$
$$+ \sum {}^{-,+}(n-1-v)(w) + \sum {}^{-,-}(n-1-v)(n-1-w)$$
$$= \sum (vw) + (n-1) \sum {}^{+,-}(v) + (n-1) \sum {}^{-,+}(w)$$
$$+ (n-1) \sum {}^{-,-}(v) + (n-1) \sum {}^{-,-}(w) + t(n-1)^2$$
$$= \sum (vw) + (n-1) \sum {}^{-w}(v) + (n-1) \sum {}^{-v}(w) + t(n-1)^2,$$

where t is the number of cells in which both the "tens" and the "units" number are negative.

Now assume that the following conditions are met and substitute these values in the equation for the sum of cd.

CONDITION 9: $\Sigma(vw) = 0$.

CONDITION 10: $\Sigma - w(v) + \Sigma - v(w) = 0$.

CONDITION 11:   One-fourth of the cells have both the "tens" number and the "units" number negative, that is, $t = n/4$. Since it is not possible for this condition and Conditions 3 and 4 to be met simultaneously without one-fourth of the cells having both signs positive, one-fourth of the cells having both signs negative, one-fourth of the cells having the "tens" number positive and the "units" number negative, and the remaining fourth of the cells having the "tens" number negative and the "units" number positive, these conditions will normally be combined and expressed in this manner.

This gives:

$$\sum(cd) = n(n-1)^2/4.$$

As a final step in checking the sufficiency of these conditions, the derived values for the sums of c^2, d^2 and cd may be substituted in the equation for the sum of q^2 as follows:

$$\sum(q^2) = n^2 \sum(c^2) + 2n \sum(cd) + \sum(d^2)$$

$$= n^3(n-1)(2n-1)/6 + 2n^2(n-1)^2/4 + n(n-1)(2n-1)/6$$

$$= n(n-1)(2n^3 - n^2 + 3n^2 - 3n + 2n - 1)/6$$

$$= n(n-1)(2n^3 + 2n^2 - n - 1)/6$$

$$= n(n^2-1)(2n^2-1)/6, \text{ as required.}$$

It follows that if Conditions 1 to 11, inclusive, are met the resultant main square will be pandiagonal in the first degree and will also be bimagic.

An analysis of the method of construction and the properties of the generating square, Figure 50, will show that, if Requirements D, E and H are met, the numbers forming the "tens" and "units" of each rcd will be one of the following combinations (the order of their appearance is immaterial):

$$+0, +1, +2, +3, -3, -2, -1, -0,$$

$$+0, -1, -2, +3, +0, -1, -2, +3,$$

or

$$-0, +1, +2, -3, -0, +1, +2, -3.$$

In each rcd, therefore, Conditions 5, 6, 7 and 8 will be met.

A similar analysis will show that Condition 9 is automatically met as far as the rows and columns are concerned. Requirement J ensures that the main diagonals meet it.

In the case of Condition 10, Requirements D and E ensure that the rows and columns will be correct (in those cases which are not met automatically) and Requirement I takes care of the main diagonals.

Likewise, in the case of Condition 11, Requirements B, C and F ensure that the rows and columns will be correct and Requirement G takes care of the main diagonals.

It is thus seen that a square constructed in accordance with Steps 1 to 4, inclusive, from the generating square in Figure 50 and a primary square which meets the Requirements A to J, inclusive, will meet Conditions 1 to 11, inclusive, and hence will be bimagic with the added property that the square will be pandiagonal in the first degree.

Let us now examine Figure 55b. The analysis is really quite simple. Examination will show that each of the capital letters A, B and C is paired once, and only once, with each of the lower-case letters a, b and c in each rcd. It follows that, when the values 0, 3 and 6 are substituted for the capital letters and 0, 1 and 2 for the lower-case letters, in each rcd the "tens" numbers will consist of the numbers 0 to 8, inclusive. The same situation holds, of course, if 0, 1 and 2 are substituted (the order being immaterial) for A, B and C, and 0, 3 and 6 for the lower-case letters a, b and c.

A similar situation exists with the letters **A**, **B** and **C**, and **a**, **b** and **c**. Therefore, the sum of the "tens" numbers and the "units" numbers, as well as the sum of their squares, will be correct for each rcd!

It remains only to show that the sum of the cross-products will be correct. Here again it is merely a case of careful analysis. A check will show that for each rcd the sum of the cross-products is

$$A(A+B+C) + B(A+B+C) + C(A+B+C)$$

$$+ A(a+b+c) + B(a+b+c) + C(a+b+c)$$

$$+ a(A+B+C) + b(A+B+C) + c(A+B+C)$$

$$+ a(a+b+c) + b(a+b+c) + c(a+b+c).$$

Collecting terms shows that this sum equals:

$$(A+B+C)(A+B+C) + (A+B+C)(a+b+c)$$

$$+ (a+b+c)(A+B+C) + (a+b+c)(a+b+c), \quad \text{giving us}$$

$$(0+3+6)(0+3+6) + (0+3+6)(0+1+2)$$

$$+ (0+1+2)(0+3+6) + (0+1+2)(0+1+2)$$

$$= 81 + 27 + 27 + 9 = 144, \quad \text{as required!}$$

29

TRIMAGIC SQUARES

Since it will result in a lot of tiresome repetition, we will not make a complete analysis of the sufficiency of the requirements listed in Tables 1 to 4, inclusive, in Chapter 13. The proof is, however, quite similar to the one employed in the last two chapters.

Assume that Table 8 is correct.

TABLE 8

$\Sigma(c^3)$	$= n^2(n-1)^2/4$
$\Sigma(d^3)$	$= n^2(n-1)^2/4$
$\Sigma(c^2d)$	$= n(2n-1)(n-1)^2/12$
$\Sigma(cd^2)$	$= n(2n-1)(n-1)^2/12$
$\Sigma(cd)$	$= n(n-1)^2/4$
$\Sigma(c^2)$	$= n(2n-1)(n-1)/6$
$\Sigma(d^2)$	$= n(2n-1)(n-1)/6$
$\Sigma(c)$	$= n(n-1)/2$
$\Sigma(d)$	$= n(n-1)/2$

Remembering that q equals $(cn+d)$ when expressed to the base n (where n is the order of the square), this gives us:

$$\sum(q) = \sum(cn+d) = n\sum(c) + \sum(d)$$

$$= n(n)(n-1)/2 + (n)(n-1)/2 = n(n^2-1)/2, \quad \text{as required.}$$

$$\sum(q^2) = \sum(cn+d)^2 = n^2\sum(c^2) + 2n\sum(cd) + \sum(d^2)$$

$$= n^2(n)(2n-1)(n-1)/6 + 2n(n)(n-1)^2/4$$

$$+ n(2n-1)(n-1)/6 = n(2n^2-1)(n^2-1)/6, \quad \text{as required.}$$

$$\sum(q^3) = \sum(cn+d)^3 = n^3\sum(c^3) + 3n^2\sum(c^2d) + 3n\sum(cd^2) + \sum(d^3)$$

$$= n^3(n^2)(n-1)^2/4 + 3n^2(n)(2n-1)(n-1)^2/12$$

$$+ 3n(n)(2n-1)(n-1)^2/12 + (n^2)(n-1)^2/4$$

$$= n^3(n^2-1)^2/4, \quad \text{as required.}$$

We thus see that if the conditions listed in Table 8 are met the final square will be trimagic. Let us now consider Table 9.

TABLE 9

$\Sigma(v^3)$	$=0$
$\Sigma(w^3)$	$=0$
$\Sigma(v^2w)$	$=0$
$\Sigma(vw^2)$	$=0$
$\Sigma+v(v^2)$	$=\Sigma-v(v^2)=\Sigma+w(v^2)=\Sigma-w(v^2)=n(n/2-1)(n-1)/12$
$\Sigma+v(w^2)$	$=\Sigma-v(w^2)=\Sigma+w(w^2)=\Sigma-w(w^2)=n(n/2-1)(n-1)/12$
$\Sigma+v(vw)$	$=\Sigma-v(vw)=\Sigma+w(vw)=\Sigma-w(vw)=0$
$\Sigma+,+(v)$	$=\Sigma+,-(v)=-\Sigma-,+(v)=-\Sigma-,-(v)=n(n/2-1)/8$
$\Sigma+,+(w)$	$=-\Sigma+,-(w)=\Sigma-,+(w)=-\Sigma-,-(w)=n(n/2-1)/8$

The number of $[+,+]$, $[+,-]$, $[-,+]$, and $[-,-]$, combinations must each equal $n/4$.

Remembering the $+v$ equals v and $-v$ is shorthand for $(n-1-$

v), we have:

$$\sum (c^3) = \sum + v(v^3) + \sum - v(n-1-v)^3$$

$$= \sum (v^3) + 3(n-1)\sum - v(v^2) + 3(n-1)^2 \sum - v(v) + (s+t)(n-1)^3,$$

where s equals the number of $[-, +]$ combinations and t equals the number of $[-, -]$ combinations.

Assuming that the conditions listed in Table 9 are met, this becomes:

$$\sum (c^3) = 0 + 3(n-1)(n)(n/2-1)(n-1)/12$$

$$+ 3(n-1)^2(-n)(n/2-1)/4 + (n/4+n/4)(n-1)^3$$

$$= n^2(n-1)^2/4, \qquad \text{as required.}$$

In a similar manner the sum of (d^3), (c^2d), (cd^2), (cd), (c^2), (d^2), (c) and (d) may be checked and it will be found that, in each instance, the values will equal those listed in Table 8 if the conditions listed in Table 9 are met. In other words we may replace the requirements for a trimagic square given in Table 8 by the longer (but simpler) requirements given by Table 9.

The next step is to determine what requirements must be met by x and y in the primary square and by a and b in the generating square in order that v and w will meet these requirements.

A careful check will show that if all of the Requirements listed in Tables 2, 3 and 4 (as the case may be) of Chapter 13 are met, the conditions listed in Table 9 above will automatically be met and the final square will be trimagic!!!

30

NONCYCLICAL PANDIAGONAL MAGIC SQUARES

In Chapter 15 we stated that the sufficient requirements which, if met, would enable you (by the careful selection of the columns) to select two pairs of columns which you could interchange and have the remaining square be a noncyclical pandiagonal magic square included (in addition to those necessary to ensure that the square would be an ordinary cyclical pandiagonal magic square) $(C^2 + R^2)$ $\equiv (c^2 + r^2) \equiv 0 \bmod n$. The proof of this statement is a bit involved, but actually it is really quite straightforward.

Let the number X be in the cell (i,j) and the number Y equal to $X + a + bn$ be in cell $(i, j+1)$. Then, since X and Y are in the same row and $Y - X = a + bn$, we have (as shown in Chapter 26)

$$ar + bR \equiv 0 \bmod n.$$

Also, since Y is the column next to X, we have

$$ac + bC \equiv 1 \bmod n.$$

Hence:

$$acr + bcR \equiv 0 \bmod n$$

and

$$acr + brC \equiv r \bmod n,$$

or

$$brC - bcR \equiv r \bmod n.$$

This gives:

$$b \equiv r/(Cr - cR) \bmod n = v \qquad \text{(for convenience)}.$$

Similarly, for X in (i,j) and Y in $(i+1,j)$ we have

$$ar + bR \equiv 1 \bmod n$$

and

$$ac + bC \equiv 0 \bmod n.$$

Hence:

$$acr + bcR \equiv c \bmod n$$

and

$$acr + brC \equiv 0 \bmod n,$$

or

$$bcR - brC \equiv c \bmod n.$$

This gives:

$$b \equiv -c/(Cr - cR) \bmod n = w \qquad \text{(for convenience)}.$$

To interchange two pairs of columns symmetrically located about a cell arbitrarily selected at random and (1) retain the regular pandiagonal property and (2) lose the cyclical property of the rows and diagonals, it is sufficient for Figure 91 to be correct, when the distances between the various rows and columns involved are as indicated.

Remember that, for a square constructed for a given set of values of C, R, c and r, the value of b to move one column to the right is $v = r/(Cr - cR) \bmod n$ and to move one row up is $w = -c/(Cr -$

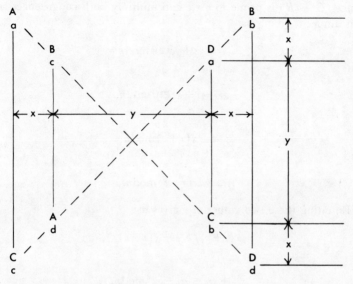

Fig. 91.

$cR)\bmod n$. Thus, if we are to go from the cell occupied by $[A,a]$ to that occupied by $[A,d]$ we must go x columns to the right and $x+y$ rows down. Also, since the value of the capital letter is not changed we have:

$$x(v)-(x+y)w\equiv 0\bmod n.$$

Similarly, as we go from $[B,b]$ to $[B,c]$ the value of B does not change during the travel of $x+y$ columns to the left and x rows down, or:

$$-(x+y)v-x(w)\equiv 0\bmod n.$$

Substituting the above values for v and w in these two congruences gives:

$$\frac{xr}{(Cr-cR)}-\frac{(x+y)(-c)}{(Cr-cR)}\equiv 0\bmod n$$

and

$$\frac{-(x+y)r}{(Cr-cR)}-\frac{x(-c)}{(Cr-cR)}\equiv 0\bmod n.$$

Since $(Cr - cR)$ is prime to n we can multiply both congruences by this amount, giving us:

$$xr + xc + yc \equiv 0 \bmod n$$

and

$$-xr - yr + xc \equiv 0 \bmod n,$$

or

$$cy \equiv -x(r + c) \bmod n$$

and

$$ry \equiv -x(r - c) \bmod n.$$

Equating these two values of y gives us:

$$-x(r + c)/c \equiv -x(r - c)/r,$$

or

$$xr^2 + xrc \equiv xrc - xc^2 \bmod n,$$

or

$$x(r^2 + c^2) \equiv 0 \bmod n.$$

Treating the lower-case letters in a similar manner will show that the necessary requirement is that $(R^2 + C^2) \equiv 0 \bmod n$, as stated previously.

Further extension of the same method will show that if these requirements are met, instead of transposing the columns to get the desired noncyclical regular pandiagonal magic square, it will be possible to transpose either four rows (thus making the columns and diagonals noncyclical) or four parallel diagonals (thus making the other diagonals, the rows and diagonals noncyclical).

The requirements which must be met in order to transfer three pairs of columns or rows have been investigated (using the same approach as in the case of the transposition of two pairs) and found to be: $(C^2 + 3R^2) \equiv 0 \bmod n$ and $(c^2 + 3r^2) \equiv 0 \bmod n$. To interchange three pairs of parallel diagonals analysis will show that $(C^3 + R^3) \equiv 0 \bmod n$ and $(c^3 + r^3) \equiv 0 \bmod n$ are necessary.

APPENDIX

COMPLETE LISTING OF ALL POSSIBLE
FOURTH-ORDER MAGIC SQUARES

The following pages present, in tabular form, the possible fourth-order magic squares—880 in all—as discussed in Chapter 18.

The numeral in parentheses above each square (to the left) is the Frénicle index number. The Roman numeral above each square (to the right) indicates the type (as defined by Dudeney) to which the square belongs. As a matter of convenience, Figure 67 (which lists the number of squares of each type) is repeated below.

CENSUS OF FOURTH-ORDER MAGIC SQUARES

Pandiagonal	Type	I		48
Semipandiagonal	Type	II	48	
	Type	III	48	
	Type	IV	96	
	Type	V	96	
	Type	VI	96	
	Total			384
Simple	Type	VI	208	
	Type	VII	56	
	Type	VIII	56	
	Type	IX	56	
	Type	X	56	
	Type	XI	8	
	Type	XII	8	
	Total			448
All Types	Total			880

```
( 1)   VI        ( 2)   VI        ( 3)  XII        ( 4)   VI        ( 5)   VI        ( 6)   VI        ( 7)   VI        ( 8) VIII
 1  2 15 16       1  2 15 16       1  2 16 15       1  3 14 16       1  3 14 16       1  3 14 16       1  3 14 16       1  3 16 14
12 14  3  5      13 14  3  4      13 14  4  3      10 13  4  7      12 13  4  5      15 13  4  2      15 13  4  2       8 15  2  9
13  7 10  4      12  7 10  5      12  7  9  6      15  6 11  2      15  8  9  2      10  6 11  7      12  8  9  5      13  6 11  4
 8 11  6  9       8 11  6  9       8 11  5 10       8 12  5  9       6 10  7 11       8 12  5  9       6 10  7 11      12 10  5  7

( 9) VIII        (10) VIII        (11) VIII        (12)   VI        (13)   VI        (14)   VI        (15)   VI        (16)   VI
 1  3 16 14       1  3 16 14       1  3 16 14       1  4 13 16       1  4 13 16       1  4 13 16       1  4 13 16       1  4 13 16
12 15  2  5      13 15  2  4      13 15  2  4       8 14  3  9       8 15  2  9      12 14  3  5      12 15  2  5      14 15  2  3
13 10  7  4       8  6 11  9      12 10  7  5      15  5 12  2      14  5 12  3      15  9  8  2      14  9  8  3       8  5 12  9
 8  6  9 11      12 10  5  7       8  6  9 11      10 11  6  7      11 10  7  6       6  7 10 11       7  6 11 10      11 10  7  6

(17)   VI        (18)   VI        (19)   VI        (20) VIII        (21)   II        (22)   II        (23)   IX        (24)   IV
 1  4 13 16       1  4 13 16       1  4 13 16       1  4 14 15       1  4 14 15       1  4 14 15       1  4 14 15       1  4 14 15
14 15  2  3      15 14  3  2      15 14  3  2       9 12  6  7      13 16  2  3      13 16  2  3      16 11  5  2      16 13  3  2
12  9  8  5       8  5 12  9      12  9  8  5      16  5 11  2       8  5 11 10      12  9  7  6       9  6 12  7       7  6 12  9
 7  6 11 10      10 11  6  7       6  7 10 11       8 13  3 10      12  9  7  6       8  5 11 10       8 13  3 10      10 11  5  8

(25)   IV        (26) VIII        (27)   II        (28)   II        (29)   IX        (30)   IV        (31)   IV        (32)    V
 1  4 14 15       1  4 15 14       1  4 15 14       1  4 15 14       1  4 15 14       1  4 15 14       1  4 15 14       1  4 16 13
16 13  3  2       9 12  7  6      13 16  3  2      13 16  3  2      16 10  5  3      16 13  2  3      16 13  2  3      14 15  3  2
11 10  8  5      16  5 10  3       8  5 10 11      12  9  6  7       9  7 12  6       6  7 12  9      10 11  8  5       7  6 10 11
 6  7  9 12       8 13  2 11      12  9  6  7       8  5 10 11       8 13  2 11      11 10  5  8       7  6  9 12      12  9  5  8

(33)    V        (34)    V        (35)    V        (36)   VI        (37)   VI        (38)   VI        (39)   VI        (40) VIII
 1  4 16 13       1  4 16 13       1  4 16 13       1  5 12 16       1  5 12 16       1  5 12 16       1  5 12 16       1  5 16 12
14 15  3  2      15 14  2  3      15 14  2  3      10 11  6  7      14 11  6  3      15 11  6  2      15 11  6  2       8 14  3  9
11 10  6  7       6  7 11 10      10 11  7  6      15  4 13  2      15  8  9  2      10  4 13  7      14  8  9  3      10  4 13  7
 8  5  9 12      12  9  5  8       8  5  9 12       8 14  3  9       4 10  7 13       8 14  3  9       4 10  7 13      15 11  2  6

(41) VIII        (42) VIII        (43) VIII        (44)   VI        (45)   VI        (46)   VI        (47)   VI        (48)   VI
 1  5 16 12       1  5 16 12       1  5 16 12       1  6 11 16       1  6 11 16       1  6 11 16       1  6 11 16       1  6 11 16
10 14  3  7      10 14  3  7      15 14  3  2       7 15  2 10       8 12  5  9       8 15  2  9      12 10  7  5      12 15  2  5
 8  4 13  9      15 11  6  2      10 11  6  7      14  4 13  3      15  3 14  2      12  3 14  5      13  3 14  4       8  3 14  9
15 11  2  6       8  4  9 13       8  4  9 13      12  9  8  5      10 13  4  7      13 10  7  4       8 15  2  9      13 10  7  4

(49)   VI        (50)   VI        (51)   VI        (52)   VI        (53)   VI        (54)   VI        (55)   VI        (56)   II
 1  6 11 16       1  6 11 16       1  6 11 16       1  6 11 16       1  6 11 16       1  6 11 16       1  6 11 16       1  6 12 15
12 15  2  5      13 10  7  4      14 12  5  3      14 15  2  3      14 15  2  3      15 12  5  2      15 12  5  2      11 16  2  5
14  9  8  3      12  3 14  5      15  9  8  2       7  4 13 10      12  9  8  5       8  3 14  9      14  9  8  3       8  3 13 10
 7  4 13 10       8 15  2  9       4  7 10 13      12  9  8  5       7  4 13 10      10 13  4  7       4  7 10 13      14  9  7  4

(57)   II        (58) VIII        (59)   IX        (60)   IV        (61)   IV        (62)   II        (63)   II        (64)   IV
 1  6 12 15       1  6 12 15       1  6 12 15       1  6 12 15       1  6 12 15       1  6 15 12       1  6 15 12       1  6 15 12
11 16  2  5      13 10  8  3      16  9  7  2      16 11  5  2      16 11  5  2      11 16  5  2      11 16  5  2      16 11  2  5
14  9  7  4      16  7  9  2      13  8 10  3       7  4 14  9      13 10  8  3       8  3 10 13      14  9  4  7       4  7 14  9
 8  3 13 10       4 11  5 14       4 11  5 14      10 13  3  8       4  7  9 14      14  9  4  7       8  3 10 13      13 10  3  8

(65)   IV        (66)    V        (67)    V        (68)    V        (69)    V        (70)   VI        (71)   VI        (72)   VI
 1  6 15 12       1  6 16 11       1  6 16 11       1  6 16 11       1  6 16 11       1  7 10 16       1  7 10 16       1  7 10 16
16 11  2  5      12 15  5  2      12 15  5  2      15 12  2  5      15 12  2  5       8 12  5  9       8 14  3  9      12  9  8  5
10 13  8  3       7  4 10 13      13 10  4  7       4  7 13 10      10 13  7  4      14  2 15  3      12  2 15  5      15  4 13  2
 7  4  9 14      14  9  3  8       8  3  9 14      14  9  3  8       8  3  9 14      11 13  4  6      13 11  6  4       6 14  3 11

(73)   VI        (74)   VI        (75)   VI        (76)   VI        (77)   VI        (78)   VI        (79)   VI        (80)   VI
 1  7 10 16       1  7 10 16       1  7 10 16       1  7 10 16       1  7 10 16       1  7 10 16       1  7 10 16       1  7 10 16
12 14  3  5      12 14  3  5      14  9  8  3      14 12  5  3      14 12  5  3      15  9  8  2      15  9  8  2      15 12  5  2
 8  2 15  9      15  9  8  2      15  6 11  2       8  2 15  9      15  9  8  2      12  4 13  5      14  6 11  3      14  9  8  3
13 11  6  4       6  4 13 11       4 12  5 13      11 13  4  6       4  6 11 13       6 14  3 11       4 12  5 13       4  6 11 13
```

```
( 81)   VI      ( 82)   II      ( 83)   II      ( 84)   IV      ( 85)   IV      ( 86)   VI      ( 87)   VI      ( 88)  XII
 1  7 10 16      1  7 12 14      1  7 12 14      1  7 12 14      1  7 12 14      1  7 14 12      1  7 14 12      1  7 14 12
15 14  3  2     10 16  3  5     10 16  3  5     16 10  5  3     16 10  5  3      8 13  2 11      9 15  4  6      9 15  4  6
12  9  8  5      8  2 13 11     15  9  6  4      6  4 15  9     13 11  8  2      9  4 15  6      8  2 13 11     16 10  5  3
 6  4 13 11     15  9  6  4      8  2 13 11     11 13  2  8      4  6  9 15     16 10  3  5     16 10  3  5      8  2 11 13

( 89)   II      ( 90)   II      ( 91)   VI      ( 92)   IV      ( 93)   IV      ( 94) VIII      ( 95)    V      ( 96)    V
 1  7 14 12      1  7 14 12      1  7 14 12      1  7 14 12      1  7 14 12      1  7 16 10      1  7 16 10      1  7 16 10
10 16  5  3     10 16  5  3     11 13  2  8     16 10  3  5     16 10  3  5     11 13  4  6     12 14  5  3     12 14  5  3
 8  2 11 13     15  9  4  6      6  4 15  9      4  6 15  9     11 13  8  2     14 12  5  3      6  4 11 13     13 11  4  6
15  9  4  6      8  2 11 13     16 10  3  5     13 11  2  8      6  4  9 15      8  2  9 15     15  9  2  8      8  2  9 15

( 97)    V      ( 98)    V      ( 99) VIII      (100)   VI      (101)    V      (102)    I      (103)    V      (104)    I
 1  7 16 10      1  7 16 10      1  7 16 10      1  8  9 16      1  8 10 15      1  8 10 15      1  8 10 15      1  8 10 15
14 12  3  5     14 12  3  5     14 13  4  3     14 13  4  3     11 14  4  5     12 13  3  6     13 12  6  3     14 11  5  4
 4  6 13 11     11 13  6  4     11 12  5  6      7  2 15 10     16  9  7  2      7  2 16  9     16  9  7  2      7  2 16  9
15  9  2  8      8  2  9 15      8  2  9 15     12 11  6  5      6  3 13 12     14 11  5  4      4  5 11 14     12 13  3  6

(105)   IX      (106)    V      (107)    I      (108)    V      (109)    I      (110)   VI      (111)   VI      (112)  III
 1  8 10 15      1  8 11 14      1  8 11 14      1  8 11 14      1  8 11 14      1  8 12 13      1  8 12 13      1  8 12 13
16 13  3  2     10 15  4  5     12 13  2  7     13 12  7  2     15 10  5  4     10 15  3  6     11 14  2  7     14 11  7  2
 5  4 14 11     16  9  6  3      6  3 16  9     16  9  6  3      6  3 16  9      7  2 14 11      6  3 15 10     15 10  6  3
12  9  7  6      7  2 13 12     15 10  5  4      4  5 10 15     12 13  2  7     16  9  5  4     16  9  5  4      4  5  9 16

(113)  III      (114)    V      (115)    V      (116)    I      (117)    I      (118)   IX      (119)   VI      (120)  III
 1  8 12 13      1  8 13 12      1  8 13 12      1  8 13 12      1  8 13 12      1  8 13 12      1  8 14 11      1  8 14 11
15 10  6  3     10 15  6  3     11 14  7  2     14 11  2  7     15 10  3  6     16 11  2  5     10 15  5  4     12 13  7  2
14 11  7  2     16  9  4  5     16  9  4  5      4  5 16  9      4  5 16  9      3  6 15 10      7  2 12 13     15 10  4  5
 4  5  9 16      7  2 11 14      6  3 10 15     15 10  3  6     14 11  2  7     14  9  4  7     16  9  3  6      6  3  9 16

(121)   VI      (122)  III      (123)   VI      (124)  III      (125)   VI      (126)  III      (127)   VI      (128) VIII
 1  8 14 11      1  8 14 11      1  8 15 10      1  8 15 10      1  8 15 10      1  8 15 10      1  9  8 16      1  9 16  8
13 12  2  7     15 10  4  5     11 14  5  4     12 13  6  3     13 12  3  6     14 11  4  5     14 15  2  3     14 12  5  3
 4  5 15 10     12 13  7  2      6  3 12 13     14 11  4  5      4  5 14 11     12 13  6  3      7  4 13 10      4  6 11 13
16  9  3  6      6  3  9 16     16  9  2  7      7  2  9 16     16  9  2  7      7  2  9 16     12  6 11  5     15  7  2 10

(129) VIII      (130)   VI      (131)   VI      (132)   VI      (133)   VI      (134)   VI      (135)   VI      (136)   VI
 1  9 16  8      1 10  7 16      1 10  7 16      1 10  7 16      1 10  7 16      1 10  7 16      1 10  7 16      1 10  7 16
15 12  5  2     12  8  9  5     12 13  4  5     12 15  2  5     14  8  9  3     14 11  6  3     14 15  2  3     15  8  9  2
 4  7 10 13     15  3 14  2     15  8  9  2      8  3 14  9     15  5 12  2     15  8  9  2      8  5 12  9     12  3 14  5
14  6  3 11      6 13  4 11      6  3 14 11     13  6 11  4      4 11  6 13      4  5 12 13     11  4 13  6      6 13  4 11

(137)   VI      (138)   VI      (139)   VI      (140)   IV      (141)   IV      (142)   IX      (143)   VI      (144)   VI
 1 10  7 16      1 10  7 16      1 10  7 16      1 10  8 15      1 10  8 15      1 10  8 15      1 10 15  8      1 10 15  8
15  8  9  2     15 11  6  2     15 13  4  2     16  7  9  2     16  7  9  2     16 13  3  2     12 13  6  3     14 11  4  5
14  5 12  3     14  8  9  3     12  8  9  5     11  4 14  5     13  6 12  3      5  4 14 11      5  4 11 14      3  6 13 12
 4 11  6 13      4  5 12 13      6  3 14 11      6 13  3 12      4 11  5 14     12  7  9  6     16  7  2  9     16  7  2  9

(145)   IX      (146)   IV      (147)   IV      (148)    V      (149)    V      (150)   VI      (151)   VI      (152)   VI
 1 10 15  8      1 10 15  8      1 10 15  8      1 10 16  7      1 10 16  7      1 11  6 16      1 11  6 16      1 11  6 16
16  6  3  9     16  7  2  9     16  7  2  9     15  8  2  9     15  8  2  9     12  8  9  5     12 14  3  5     12 14  3  5
 5 11 14  4      4 11 14  5      6 13 12  3      4 11 13  6      6 13 11  4     14  2 15  3      8  2 15  9     13  7 10  4
12  7  2 13     13  6  3 12     11  4  5 14     14  5  3 12     12  3  5 14      7 13  4 10     13  7 10  4      8  2 15  9

(153)   VI      (154)   VI      (155)   VI      (156)   VI      (157)   VI      (158)   VI      (159)   IV      (160)   IV
 1 11  6 16      1 11  6 16      1 11  6 16      1 11  6 16      1 11  6 16      1 11  6 16      1 11  8 14      1 11  8 14
13 14  3  4     14  8  9  3     14  8  9  3     14 13  4  3     15  8  9  2     15 14  3  2     16  6  9  3     16  6  9  3
12  7 10  5     12  2 15  5     15  5 12  2      7  2 15 10     14  5 12  3      8  5 12  9     10  4 15  5     13  7 12  2
 8  2 15  9      7 13  4 10      4 10  7 13     12  8  9  5      4 10  7 13     10  4 13  7      7 13  2 12      4 10  5 15
```

```
(161)   IX      (162)   IV      (163)   IV      (164)    V      (165)    V      (166)   VI      (167)   VI      (168)   VI
 1 11 14  8      1 11 14  8      1 11 14  8      1 11 16  6      1 11 16  6      1 12  5 16      1 12  5 16      1 12  5 16
16  5  4  9     16  6  3  9     16  6  3  9     14  8  3  9     14  8  3  9     14  9  8  3     15  9  8  2     15 13  4  2
 7 12 13  2      4 10 15  5      7 13 12  2      4 10 13  7      7 13 10  4     15  6 11  2     14  6 11  3     10  6 11  7
10  6  3 15     13  7  2 12     10  4  5 15     15  5  2 12     12  2  5 15      4  7 10 13      4  7 10 13      8  3 14  9

(169)    V      (170) VIII      (171)    I      (172)   IX      (173)    V      (174)    I      (175)  III      (176)  III
 1 12  6 15      1 12  6 15      1 12  6 15      1 12  6 15      1 12  7 14      1 12  7 14      1 12  8 13      1 12  8 13
13  8 10  3     13 10  8  3     14  7  9  4     16  9  7  2     13  8 11  2     15  6  9  4     14  7 11  2     15  6 10  3
16  5 11  2     16  7  9  2     11  2 16  5     13  8 10  3     16  5 10  3     10  3 16  5     15  6 10  3     14  7 11  2
 4  9  7 14      4  5 11 14      8 13  3 10      4  5 11 14      4  9  6 15      8 13  2 11      4  9  5 16      4  9  5 16

(177)    I      (178)    I      (179)   VI      (180)   IX      (181)   XI      (182)   VI      (183)  III      (184)   VI
 1 12 13  8      1 12 13  8      1 12 13  8      1 12 13  8      1 12 13  8      1 12 14  7      1 12 14  7      1 12 15  6
14  7  2 11     15  6  3 10     15 10  3  6     16  7  2  9     16  9  4  5     13  8  2 11     15  6  4  9     13  8  3 10
 4  9 16  5      4  9 16  5      2  7 14 11      3 10 15  6      2  7 14 11      4  9 15  6      8 13 11  2      4  9 14  7
15  6  3 10     14  7  2 11     16  5  4  9     14  5  4 11     15  6  3 10     16  5  3 10     10  3  5 16     16  5  2 11

(185)  III      (186)   VI      (187)   VI      (188)   VI      (189)   VI      (190)   VI      (191)   IV      (192)   IV
 1 12 15  6      1 12 15  6      1 13  4 16      1 13  4 16      1 13  4 16      1 13  4 16      1 13  8 12      1 13  8 12
14  7  4  9     14  9  4  7     14  8  9  3     14 12  5  3     15  8  9  2     15 12  5  2     16  4  9  5     16  4  9  5
 8 13 10  3      3  8 13 10     12  2 15  5      8  2 15  9     12  3 14  5      8  3 14  9     10  6 15  3     11  7 14  2
11  2  5 16     16  5  2 11      7 11  6 10     11  7 10  6      6 10  7 11     10  6 11  7      7 11  2 14      6 10  3 15

(193)   IX      (194)   VI      (195)   IV      (196)   IV      (197)   IX      (198)   VI      (199)   VI      (200)   IX
 1 13  8 12      1 13 12  8      1 13 12  8      1 13 12  8      1 13 12  8      1 14  3 16      1 14  3 16      1 14  4 15
16 11  2  5     15 10  3  6     16  4  5  9     16  4  5  9     16  7  2  9     15  9  8  2     15 11  6  2     16 11  5  2
 3  6 15 10      2  7 14 11      6 10 15  3      7 11 14  2      3 10 15  6     12  4 13  5     10  4 13  7      9  6 12  7
14  4  9  7     16  4  5  9     11  7  2 14     10  6  3 15     14  4  5 11      6  7 10 11      8  5 12  9      8  3 13 10

(201)    I      (202)   XI      (203)  III      (204)    I      (205)   IX      (206)  III      (207)   IX      (208)   IX
 1 14  7 12      1 14  7 12      1 14  8 11      1 14 11  8      1 14 11  8      1 14 12  7      1 15  4 14      1 15 10  8
15  4  9  6     16  5 10  3     15  4 10  5     15  4  5 10     16  5  4  9     15  4  6  9     16 10  5  3     16  6  3  9
10  5 16  3      9  4 15  6     12  7 13  2      6  9 16  3      7 12 13  2      8 11 13  2      9  7 12  6      5 11 14  4
 8 11  2 13      8 11  2 13      6  9  3 16     12  7  2 13     10  3  6 15     10  5  3 16      8  2 13 11     12  2  7 13

(209)  XII      (210)   VI      (211)   VI      (212) VIII      (213)   II      (214)   II      (215)   IX      (216)   IV
 2  1 15 16      2  1 16 15      2  1 16 15      2  3 13 16      2  3 13 16      2  3 13 16      2  3 13 16      2  3 13 16
14 13  3  4     11 13  4  6     14 13  4  3     10 11  5  8     14 15  1  4     14 15  1  4     15 12  6  1     15 14  4  1
11  8 10  5     14  8  9  3     11  8  9  6     15  6 12  1      7  6 12  9     11 10  8  5     10  5 11  8      8  5 11 10
 7 12  6  9      7 12  5 10      7 12  5 10      7 14  4  9     11 10  8  5      7  6 12  9      7 14  4  9      9 12  6  7

(217)   IV      (218)   VI      (219)   VI      (220)   VI      (221)   VI      (222)   VI      (223)   VI      (224)   VI
 2  3 13 16      2  3 14 15      2  3 14 15      2  3 14 15      2  3 14 15      2  3 14 15      2  3 14 15      2  3 14 15
15 14  4  1      7 13  4 10      7 16  1 10      8 16  1  9     11 13  4  6     11 16  1  6     11 16  1  6     13 16  1  4
12  9  7  6     16  6 11  1     13  6 11  4     11  5 12  6     16 10  7  1      8  5 12  9     13 10  7  4      7  6 11 10
 5  8 10 11      9 12  5  8     12  9  8  5     13 10  7  4      5  8  9 12     13 10  7  4      8  5 12  9     12  9  8  5

(225)   VI      (226)   VI      (227)   VI      (228)    V      (229)    V      (230)    V      (231)    V      (232) VIII
 2  3 14 15      2  3 14 15      2  3 14 15      2  3 15 14      2  3 15 14      2  3 15 14      2  3 15 14      2  3 16 13
13 16  1  4     16 13  4  1     16 13  4  1     13 16  4  1     13 16  4  1     16 13  1  4     16 13  1  4     10 11  8  5
11 10  7  6      7  6 11 10     11 10  7  6      8  5  9 12     12  9  5  8      5  8 12  9      9 12  8  5     15  6  9  4
 8  5 12  9      9 12  5  8      5  8  9 12     11 10  6  7      7  6 10 11     11 10  6  7      7  6 10 11      7 14  1 12

(233)   II      (234)   II      (235)   IX      (236)   IV      (237)   IV      (238)    X      (239)    X      (240)    X
 2  3 16 13      2  3 16 13      2  3 16 13      2  3 16 13      2  3 16 13      2  4 13 15      2  4 13 15      2  4 13 15
14 15  4  1     14 15  4  1     15  9  6  4     15 14  1  4     15 14  1  4     14 16  3  1     14 16  3  1     16 14  1  3
 7  6  9 12     11 10  5  8     10  8 11  5      5  8 11 10      9 12  7  6      7  5 10 12     11  9  6  8      5  7 12 10
11 10  5  8      7  6  9 12      7 14  1 12     12  9  6  7      8  5 10 11     11  9  8  6      7  5 12 10     11  9  8  6
```

(241) X
2 4 13 15
16 14 1 3
9 11 8 6
7 5 12 10

(242) VIII
2 4 15 13
5 14 3 12
16 7 10 1
11 9 6 8

(243) VIII
2 4 15 13
9 14 3 8
16 11 6 1
7 5 10 12

(244) VIII
2 4 15 13
16 14 3 1
5 7 10 12
11 9 6 8

(245) VIII
2 4 15 13
16 14 3 1
9 11 6 8
7 5 10 12

(246) II
2 5 11 16
12 15 1 6
7 4 14 9
13 10 8 3

(247) II
2 5 11 16
12 15 1 6
13 10 8 3
7 4 14 9

(248) VIII
2 5 11 16
14 9 7 4
15 8 10 1
3 12 6 13

(249) IX
2 5 11 16
15 10 8 1
14 7 9 4
3 12 6 13

(250) IV
2 5 11 16
15 12 6 1
8 3 13 10
9 14 4 7

(251) IV
2 5 11 16
15 12 6 1
14 9 7 4
3 8 10 13

(252) VI
2 5 12 15
7 11 6 10
16 4 13 1
9 14 3 8

(253) VI
2 5 12 15
7 16 1 10
11 4 13 6
14 9 8 3

(254) VI
2 5 12 15
11 9 8 6
14 4 13 3
7 16 1 10

(255) VI
2 5 12 15
11 16 1 6
7 4 13 10
14 9 8 3

(256) VI
2 5 12 15
11 16 1 6
13 10 7 4
8 3 14 9

(257) VI
2 5 12 15
13 11 6 4
16 10 7 1
3 8 9 14

(258) VI
2 5 12 15
13 16 1 4
11 10 7 6
8 3 14 9

(259) VI
2 5 12 15
14 9 8 3
11 4 13 6
7 16 1 10

(260) X
2 5 12 15
14 16 3 1
11 9 6 8
7 4 13 10

(261) VI
2 5 12 15
16 11 6 1
7 4 13 10
9 14 3 8

(262) VI
2 5 12 15
16 11 6 1
13 10 7 4
3 8 9 14

(263) X
2 5 12 15
16 14 1 3
9 11 8 6
7 4 13 10

(264) V
2 5 15 12
11 16 6 1
8 3 9 14
13 10 4 7

(265) V
2 5 15 12
11 16 6 1
14 9 3 8
7 4 10 13

(266) V
2 5 15 12
16 11 1 6
3 8 14 9
13 10 4 7

(267) V
2 5 15 12
16 11 1 6
9 14 8 3
7 4 10 13

(268) VII
2 5 16 11
8 12 1 13
9 7 14 4
15 10 3 6

(269) II
2 5 16 11
12 15 6 1
7 4 9 14
13 10 3 8

(270) II
2 5 16 11
12 15 6 1
13 10 3 8
7 4 9 14

(271) VII
2 5 16 11
13 12 1 8
4 7 14 9
15 10 3 6

(272) IX
2 5 16 11
15 10 3 6
4 7 14 9
13 12 1 8

(273) IV
2 5 16 11
15 12 1 6
3 8 13 10
14 9 4 7

(274) IV
2 5 16 11
15 12 1 6
9 14 7 4
8 3 10 13

(275) VIII
2 6 15 11
7 13 4 10
9 3 14 8
16 12 1 5

(276) VIII
2 6 15 11
9 13 4 8
7 3 14 10
16 12 1 5

(277) VIII
2 6 15 11
9 13 4 8
16 12 5 1
7 3 10 14

(278) VIII
2 6 15 11
16 13 4 1
9 12 5 8
7 3 10 14

(279) I
2 7 9 16
11 14 4 5
8 1 15 10
13 12 6 3

(280) V
2 7 9 16
12 13 3 6
15 10 8 1
5 4 14 11

(281) I
2 7 9 16
13 12 6 3
8 1 15 10
11 14 4 5

(282) V
2 7 9 16
14 11 5 4
15 10 8 1
3 6 12 13

(283) IX
2 7 9 16
15 14 4 1
6 3 13 12
11 10 8 5

(284) VI
2 7 10 15
8 12 5 9
11 1 16 6
13 14 3 4

(285) VI
2 7 10 15
11 12 5 6
8 1 16 9
13 14 3 4

(286) VI
2 7 11 14
8 13 1 12
9 4 16 5
15 10 6 3

(287) VI
2 7 11 14
9 16 4 5
8 1 13 12
15 10 6 3

(288) VI
2 7 11 14
12 13 1 8
5 4 16 9
15 10 6 3

(289) III
2 7 11 14
13 12 8 1
16 9 5 4
3 6 10 15

(290) III
2 7 11 14
16 9 5 4
13 12 8 1
3 6 10 15

(291) V
2 7 12 13
9 16 3 6
15 10 5 4
8 1 14 11

(292) I
2 7 12 13
11 14 1 8
5 4 15 10
16 9 6 3

(293) V
2 7 12 13
14 11 8 1
15 10 5 4
3 6 9 16

(294) I
2 7 12 13
16 9 6 3
5 4 15 10
11 14 1 8

(295) VI
2 7 13 12
8 11 1 14
9 6 16 3
15 10 4 5

(296) VI
2 7 13 12
9 16 6 3
8 1 11 14
15 10 4 5

(297) III
2 7 13 12
11 14 8 1
16 9 3 6
5 4 10 15

(298) VI
2 7 13 12
14 11 1 8
3 6 16 9
15 10 4 5

(299) III
2 7 13 12
16 9 3 6
11 14 8 1
5 4 10 15

(300) VII
2 7 14 11
8 10 3 13
9 5 16 4
15 12 1 6

(301) V
2 7 14 11
9 16 5 4
15 10 3 6
8 1 12 13

(302) V
2 7 14 11
12 13 8 1
15 10 3 6
5 4 9 16

(303) VII
2 7 14 11
13 10 3 8
4 5 16 9
15 12 1 6

(304) I
2 7 14 11
13 12 1 8
3 6 15 10
16 9 4 5

(305) I
2 7 14 11
16 9 4 5
3 6 15 10
13 12 1 8

(306) III
2 7 16 9
11 14 5 4
13 12 3 6
8 1 10 15

(307) VI
2 7 16 9
12 13 6 3
5 4 11 14
15 10 1 8

(308) III
2 7 16 9
13 12 3 6
11 14 5 4
8 1 10 15

(309) VI
2 7 16 9
14 11 4 5
3 6 13 12
15 10 1 8

(310) VI
2 8 9 15
11 13 4 6
7 1 16 10
14 12 5 3

(311) VI
2 8 9 15
11 13 4 6
16 10 7 1
5 3 14 12

(312) VI
2 8 9 15
13 11 6 4
7 1 16 10
12 14 3 5

(313) VI
2 8 9 15
13 11 6 4
16 10 7 1
3 5 12 14

(314) VI
2 8 9 15
16 11 6 1
13 10 7 4
3 5 12 14

(315) VI
2 8 9 15
16 13 4 1
11 10 7 6
5 3 14 12

(316) II
2 8 11 13
9 15 4 6
7 1 14 12
16 10 5 3

(317) II
2 8 11 13
9 15 4 6
16 10 5 3
7 1 14 12

(318) VI
2 8 11 13
10 16 5 3
7 1 12 14
15 9 6 4

(319) XII
2 8 11 13
10 16 5 3
15 9 4 6
7 1 14 12

(320) VI
2 8 11 13
14 12 1 7
3 5 16 10
15 9 6 4

```
(321)   IV     (322)   IV     (323)   II     (324)   II     (325)   IV     (326)   IV     (327) VIII    (328)    V
 2  8 11 13     2  8 11 13     2  8 13 11     2  8 13 11     2  8 13 11     2  8 13 11     2  8 15  9     2  8 15  9
15  9  6  4    15  9  6  4     9 15  6  4     9 15  6  4    15  9  4  6    15  9  4  6    11 12  5  6    11 13  6  4
 5  3 16 10    14 12  7  1     7  1 12 14    16 10  3  5     3  5 16 10    12 14  7  1    14 13  4  3     5  3 12 14
12 14  1  7     3  5 10 16    16 10  3  5     7  1 12 14    14 12  1  7     5  3 10 16     7  1 10 16    16 10  1  7

(329)    V     (330)    V     (331)    V     (332) VIII    (333)   IV     (334)   IV     (335)   IX     (336)   VI
 2  8 15  9     2  8 15  9     2  8 15  9     2  8 15  9     2  9  7 16     2  9  7 16     2  9  7 16     2  9  8 15
11 13  6  4    13 11  4  6    13 11  4  6    14 12  5  3    15  8 10  1    15  8 10  1    15 14  4  1    11  7 10  6
14 12  3  5     3  5 14 12    12 14  5  3    11 13  4  6    12  3 13  6    14  5 11  4     6  3 13 12    16  4 13  1
 7  1 10 16    16 10  1  7     7  1 10 16     7  1 10 16     5 14  4 11     3 12  6 13    11  8 10  5     5 14  3 12

(337)   VI     (338)   VI     (339)   VI     (340)    X     (341)   VI     (342)   VI     (343)    X     (344)    V
 2  9  8 15     2  9  8 15     2  9  8 15     2  9  8 15     2  9  8 15     2  9  8 15     2  9  8 15     2  9 15  8
11 16  1  6    13  7 10  4    13 16  1  4    14 16  3  1    16  7 10  1    16  7 10  1    16 14  1  3    16  7  1 10
 7  4 13 10    16  6 11  1     7  6 11 10     7  5 10 12    11  4 13  6    13  6 11  4     5  7 12 10     3 12 14  5
14  5 12  3     3 12  5 14    12  3 14  5    11  4 13  6     5 14  3 12     3 12  5 14    11  4 13  6    13  6  4 11

(345)    V     (346)  VII     (347)  VII     (348)   IX     (349)   IX     (350)   IV     (351)   IV     (352)   VI
 2  9 15  8     2  9 16  7     2  9 16  7     2  9 16  7     2  9 16  7     2  9 16  7     2  9 16  7     2 10  7 15
16  7  1 10    12  8  1 13    13  8  1 12    15  5  4 10    15  6  3 10    15  8  1 10    15  8  1 10    11 16  1  6
 5 14 12  3     5 11 14  4     4 11 14  5     6 12 13  3     4 11 14  5     3 12 13  6     5 14 11  4     8  5 12  9
11  4  6 13    15  6  3 10    15  6  3 10    11  8  1 14    13  8  1 12    14  5  4 11    12  3  6 13    13  3 14  4

(353) VIII    (354) VIII    (355)    I     (356)    V     (357) VIII    (358)   IX     (359)   VI     (360)  III
 2 10 15  7     2 10 15  7     2 11  5 16     2 11  5 16     2 11  5 16     2 11  5 16     2 11  7 14     2 11  7 14
13 11  6  4    16 11  6  1    13  8 10  3    14  7  9  4    14  9  7  4    15 10  8  1    12 13  1  8    13  8 12  1
 3  5 12 14     3  8  9 14    12  1 15  6    15  6 12  1    15  8 10  1    14  7  9  4     5  4 16  9    16  5  9  4
16  8  1  9    13  5  4 12     7 14  4  9     3 10  8 13     3  6 12 13     3  6 12 13    15  6 10  3     3 10  6 15

(361)  III     (362)    V     (363)   VI     (364)   XI     (365)    I     (366)   VI     (367)   VI     (368)  III
 2 11  7 14     2 11  8 13     2 11  8 13     2 11  8 13     2 11  8 13     2 11 13  8     2 11 13  8     2 11 13  8
16  5  9  4    14  7 12  1    14 12  1  7    15  4  9  6    16  5 10  3    12  7  1 14    14  7  1 12    16  5  3 10
13  8 12  1    15  6  9  4     3  5 16 10    10  5 16  3     9  4 15  6     5 10 16  3     3 10 16  5     7 14 12  1
 3 10  6 15     3 10  5 16    15  6  9  4     7 14  1 12     7 14  1 12    15  6  4  9    15  6  4  9     9  4  6 15

(369)  VII     (370)   VI     (371)  VII     (372)    I     (373)   IX     (374)   XI     (375)    I     (376)   VI
 2 11 14  7     2 11 14  7     2 11 14  7     2 11 14  7     2 11 14  7     2 11 14  7     2 11 14  7     2 11 14  7
12  6  3 13    12 13  8  1    13  6  3 12    13  8  1 12    15  4  5 10    15 10  3  6    16  5  4  9    16  9  4  5
 5  9 16  4     5  4  9 16     4  9 16  5     3 10 15  6     8 13 12  1     1  8 13 12     3 10 15  6     1  8 13 12
15  8  1 10    15  6  3 10    15  8  1 10    16  5  4  9     9  6  3 16    16  5  4  9    13  8  1 12    15  6  3 10

(377)  III     (378)   VI     (379)   VI     (380)   VI     (381)   VI     (382)   VI     (383)   VI     (384)   IV
 2 11 16  5     2 11 16  5     2 12  5 15     2 12  5 15     2 12  5 15     2 12  5 15     2 12  5 15     2 12  7 13
13  8  3 10    14  7  4  9    13  7 10  4    13  7 10  4    14 13  4  3    16  7 10  1    16 13  4  1    15  5 10  4
 7 14  9  4     3 10 13  8    11  1 16  6    16  6 11  1    11  8  9  6    13  6 11  4     7  6 11 10     9  3 16  6
12  1  6 15    15  6  1 12     8 14  3  9     3  9  8 14     7  1 16 10     3  9  8 14     9  3 14  8     8 14  1 11

(385)   IV     (386)   IV     (387)   IV     (388)    V     (389)    V     (390)   IX     (391)   VI     (392)  III
 2 12  7 13     2 12 13  7     2 12 13  7     2 12 15  5     2 12 15  5     2 13  3 16     2 13  7 12     2 13  7 12
15  5 10  4    15  5  4 10    15  5  4 10    13  7  4 10    13  7  4 10    15 12  6  1    14 11  1  8    16  3  9  6
14  8 11  1     3  9 16  6     8 14 11  1     3  9 14  8     8 14  9  3    10  5 11  8     3  6 16  9    11  8 14  1
 3  9  6 16    14  8  1 11     9  3  6 16    16  6  1 11    11  1  6 16     7  4 14  9    15  4 10  5     5 10  4 15

(393)    I     (394)   VI     (395)  III     (396)    I     (397)   VI     (398)   VI     (399)   IV     (400)   IV
 2 13  8 11     2 13 11  8     2 13 11  8     2 13 12  7     2 14  3 15     2 14  3 15     2 14  7 11     2 14  7 12
16  3 10  5    14  7  1 12    16  3  5 10    16  3  6  9    16  7 10  1    16 11  6  1    15  3 10  6    15  3 10
 9  6 15  4     3 10 16  5     7 12 14  1     5 10 15  4    11  4 13  6     7  4 13 10     9  5 16  4    12  8 13
 7 12  1 14    15  4  6  9     9  6  4 15    11  8  1 14     5  9  8 12     9  5 12  8     8 12  1 13     5  9  4 1
```

(401) IV
```
 2 14 11  7
15  3  6 10
 5  9 16  4
12  8  1 13
```

(402) IV
```
 2 14 11  7
15  3  6 10
 8 12 13  1
 9  5  4 16
```

(403) IX
```
 2 14 11  7
15  4  5 10
 8 13 12  1
 9  3  6 16
```

(404) VI
```
 2 14 11  7
16  9  4  5
 1  8 13 12
15  3  6 10
```

(405) IX
```
 2 15  4 13
16  6 11  1
 9  3 14  8
 7 10  5 12
```

(406) IX
```
 2 15  4 13
16 10  7  1
 5  3 14 12
11  6  9  8
```

(407) IX
```
 2 15  6 11
16  5 12  1
 9  4 13  8
 7 10  3 14
```

(408) IX
```
 2 15 10  7
16  9  8  1
 3  6 11 14
13  4  5 12
```

(409) VIII
```
 3  1 14 16
 8 15  2  9
13  6 11  4
10 12  7  5
```

(410) VIII
```
 3  1 14 16
12 15  2  5
13 10  7  4
 6  8 11  9
```

(411) VIII
```
 3  1 14 16
13 15  2  4
 8  6 11  9
10 12  7  5
```

(412) VIII
```
 3  1 14 16
13 15  2  4
12 10  7  5
 6  8 11  9
```

(413) X
```
 3  1 16 14
13 15  4  2
 8  6  9 11
10 12  5  7
```

(414) X
```
 3  1 16 14
13 15  4  2
12 10  5  7
 6  8  9 11
```

(415) X
```
 3  1 16 14
15 13  2  4
 6  8 11  9
10 12  5  7
```

(416) X
```
 3  1 16 14
15 13  2  4
10 12  7  5
 6  8  9 11
```

(417) VIII
```
 3  2 13 16
11 10  5  8
14  7 12  1
 6 15  4  9
```

(418) IX
```
 3  2 13 16
14 12  7  1
11  5 10  8
 6 15  4  9
```

(419) IV
```
 3  2 13 16
14 15  4  1
 8  5 10 11
 9 12  7  6
```

(420) IV
```
 3  2 13 16
14 15  4  1
12  9  6  7
 5  8 11 10
```

(421) II
```
 3  2 13 16
15 14  1  4
 6  7 12  9
10 11  8  5
```

(422) II
```
 3  2 13 16
15 14  1  4
10 11  8  5
 6  7 12  9
```

(423) V
```
 3  2 14 15
13 16  4  1
 8  5  9 12
10 11  7  6
```

(424) V
```
 3  2 14 15
13 16  4  1
12  9  5  8
 6  7 11 10
```

(425) V
```
 3  2 14 15
16 13  1  4
 5  8 12  9
10 11  7  6
```

(426) V
```
 3  2 14 15
16 13  1  4
 9 12  8  5
 6  7 11 10
```

(427) VI
```
 3  2 15 14
 6 13  4 11
16  7 10  1
 9 12  5  8
```

(428) VI
```
 3  2 15 14
 6 16  1 11
13  7 10  4
12  9  8  5
```

(429) VI
```
 3  2 15 14
10 13  4  7
16 11  6  1
 5  8  9 12
```

(430) VI
```
 3  2 15 14
10 16  1  7
13 11  6  4
 8  5 12  9
```

(431) X
```
 3  2 15 14
12 16  5  1
13  9  4  8
 6  7 10 11
```

(432) VI
```
 3  2 15 14
13 16  1  4
 6  7 10 11
12  9  8  5
```

(433) VI
```
 3  2 15 14
13 16  1  4
10 11  6  7
 8  5 12  9
```

(434) X
```
 3  2 15 14
16 12  1  5
 9 13  8  4
 6  7 10 11
```

(435) VI
```
 3  2 15 14
16 13  4  1
 6  7 10 11
 9 12  5  8
```

(436) VI
```
 3  2 15 14
16 13  4  1
10 11  6  7
 5  8  9 12
```

(437) VIII
```
 3  2 16 13
 7 12  6  9
14  5 11  4
10 15  1  8
```

(438) VII
```
 3  2 16 13
 8 15  1 10
 9  6 12  7
14 11  5  4
```

(439) VII
```
 3  2 16 13
10 15  1  8
 7  6 12  9
14 11  5  4
```

(440) VIII
```
 3  2 16 13
11 10  8  5
14  7  9  4
 6 15  1 12
```

(441) IX
```
 3  2 16 13
14  9  7  4
11  8 10  5
 6 15  1 12
```

(442) IX
```
 3  2 16 13
14 11  5  4
 7  6 12  9
10 15  1  8
```

(443) IV
```
 3  2 16 13
14 15  1  4
 5  8 10 11
12  9  7  6
```

(444) IV
```
 3  2 16 13
14 15  1  4
 9 12  6  7
 8  5 11 10
```

(445) II
```
 3  2 16 13
15 14  4  1
 6  7  9 12
10 11  5  8
```

(446) II
```
 3  2 16 13
15 14  4  1
10 11  5  8
 6  7  9 12
```

(447) VI
```
 3  4 13 14
 6 16  1 11
15  9  8  2
10  5 12  7
```

(448) VI
```
 3  4 13 14
15 16  1  2
 6  9  8 11
10  5 12  7
```

(449) XII
```
 3  4 14 13
15 16  2  1
 6  9  7 12
10  5 11  8
```

(450) II
```
 3  5 10 16
12 14  1  7
13 11  8  2
 6  4 15  9
```

(451) IV
```
 3  5 10 16
14 12  7  1
 8  2 13 11
 9 15  4  6
```

(452) VI
```
 3  5 12 14
 6 10  7 11
16  4 13  1
 9 15  2  8
```

(453) VII
```
 3  5 12 14
 6 11  8  9
15  2 13  4
10 16  1  7
```

(454) VI
```
 3  5 12 14
 6 16  1 11
15  9  8  2
10  4 13  7
```

(455) VII
```
 3  5 12 14
 8  9  6 11
13  4 15  2
10 16  1  7
```

(456) VI
```
 3  5 12 14
10 16  1  7
13 11  6  4
 8  2 15  9
```

(457) VI
```
 3  5 12 14
13 16  1  4
10 11  6  7
 8  2 15  9
```

(458) VI
```
 3  5 12 14
15 16  1  2
 6  9  8 11
10  4 13  7
```

(459) VI
```
 3  5 12 14
16 10  7  1
 6  4 13 11
 9 15  2  8
```

(460) V
```
 3  5 14 12
10 16  7  1
 8  2  9 15
13 11  4  6
```

(461) V
```
 3  5 14 12
10 16  7  1
15  9  2  8
 6  4 11 13
```

(462) V
```
 3  5 14 12
16 10  1  7
 2  8 15  9
13 11  4  6
```

(463) V
```
 3  5 14 12
16 10  1  7
 9 15  8  2
 6  4 11 13
```

(464) II
```
 3  5 16 10
12 14  7  1
 6  4  9 15
13 11  2  8
```

(465) II
```
 3  5 16 10
12 14  7  1
13 11  2  8
 6  4  9 15
```

(466) IV
```
 3  5 16 10
14 12  1  7
 2  8 13 11
15  9  4  6
```

(467) IV
```
 3  5 16 10
14 12  1  7
 9 15  6  4
 8  2 11 13
```

(468) V
```
 3  6  9 16
12 13  2  7
14 11  8  1
 5  4 15 10
```

(469) I
```
 3  6  9 16
13 12  7  2
 8  1 14 11
10 15  4  5
```

(470) VII
```
 3  6 12 13
 8 11  5 10
 9  2 16  7
14 15  1  4
```

(471) V
```
 3  6 12 13
 9 16  2  7
14 11  5  4
 8  1 15 10
```

(472) VII
```
 3  6 12 13
10 11  5  8
 7  2 16  9
14 15  1  4
```

(473) I
```
 3  6 12 13
16  9  7  2
 5  4 14 11
10 15  1  8
```

(474) VI
```
 3  6 13 12
 8 10  1 15
 9  7 16  2
14 11  4  5
```

(475) VI
```
 3  6 13 12
 9 16  7  2
 8  1 10 15
14 11  4  5
```

(476) III
```
 3  6 13 12
10 15  8  1
16  9  2  7
 5  4 11 14
```

(477) VI
```
 3  6 13 12
15 10  1  8
 2  7 16  9
14 11  4  5
```

(478) III
```
 3  6 13 12
16  9  2  7
10 15  8  1
 5  4 11 14
```

(479) V
```
 3  6 15 10
 9 16  5  4
14 11  2  7
 8  1 12 13
```

(480) VI
```
 3  6 15 10
12  8  1 13
 5  9 16  4
14 11  2  7
```

(481) V
3 6 15 10
12 13 8 1
14 11 2 7
5 4 9 16

(482) VI
3 6 15 10
13 8 1 12
4 9 16 5
14 11 2 7

(483) I
3 6 15 10
13 12 1 8
2 7 14 11
16 9 4 5

(484) XI
3 6 15 10
14 7 2 11
4 9 16 5
13 12 1 8

(485) I
3 6 15 10
16 9 4 5
2 7 14 11
13 12 1 8

(486) VII
3 6 15 10
16 11 2 5
1 8 13 12
14 9 4 7

(487) III
3 6 16 9
10 15 5 4
13 12 2 7
8 1 11 14

(488) VI
3 6 16 9
12 13 7 2
5 4 10 15
14 11 1 8

(489) III
3 6 16 9
13 12 2 7
10 15 5 4
8 1 11 14

(490) VI
3 6 16 9
15 10 4 5
2 7 13 12
14 11 1 8

(491) X
3 7 10 14
12 16 5 1
6 2 11 15
13 9 8 4

(492) X
3 7 10 14
12 16 5 1
13 9 4 8
6 2 15 11

(493) X
3 7 10 14
16 12 1 5
2 6 15 11
13 9 8 4

(494) X
3 7 10 14
16 12 1 5
9 13 8 4
6 2 15 11

(495) VIII
3 7 14 10
9 12 5 8
16 13 4 1
6 2 11 15

(496) VIII
3 7 14 10
16 12 5 1
2 6 11 15
13 9 4 8

(497) VIII
3 7 14 10
16 12 5 1
9 13 4 8
6 2 11 15

(498) VI
3 8 9 14
10 13 4 7
16 11 6 1
5 2 15 12

(499) VI
3 8 9 14
13 10 7 4
6 1 16 11
12 15 2 5

(500) X
3 8 9 14
13 15 4 2
12 10 5 7
6 1 16 11

(501) X
3 8 9 14
15 13 2 4
10 12 7 5
6 1 16 11

(502) VI
3 8 9 14
16 13 4 1
10 11 6 7
5 2 15 12

(503) II
3 8 10 13
9 14 4 7
16 11 5 2
6 1 15 12

(504) IV
3 8 10 13
14 9 7 4
5 2 16 11
12 15 1 6

(505) II
3 8 13 10
9 14 7 4
6 1 12 15
16 11 2 5

(506) II
3 8 13 10
9 14 7 4
16 11 2 5
6 1 12 15

(507) IV
3 8 13 10
14 9 4 7
2 5 16 11
15 12 1 6

(508) IV
3 8 13 10
14 9 4 7
12 15 6 1
5 2 11 16

(509) IX
3 8 13 10
14 11 2 7
1 6 15 12
16 9 4 5

(510) VII
3 8 13 10
16 9 4 5
1 6 15 12
14 11 2 7

(511) V
3 8 14 9
10 13 7 4
5 2 12 15
16 11 1 6

(512) V
3 8 14 9
10 13 7 4
15 12 2 5
6 1 11 16

(513) V
3 8 14 9
13 10 4 7
2 5 15 12
16 11 1 6

(514) V
3 8 14 9
13 10 4 7
12 15 5 2
6 1 11 16

(515) IV
3 9 6 16
14 8 11 1
12 2 13 7
5 15 4 10

(516) VI
3 9 8 14
10 6 11 7
16 4 13 1
5 15 2 12

(517) VII
3 9 8 14
10 7 12 5
15 2 13 4
6 16 1 11

(518) VII
3 9 8 14
12 5 10 7
13 4 15 2
6 16 1 11

(519) X
3 9 8 14
12 16 5 1
6 2 11 15
13 7 10 4

(520) VI
3 9 8 14
13 16 1 4
6 7 10 11
12 2 15 5

(521) VI
3 9 8 14
16 6 11 1
10 4 13 7
5 15 2 12

(522) X
3 9 8 14
16 12 1 5
2 6 15 11
13 7 10 4

(523) V
3 9 14 8
16 6 1 11
2 12 15 5
13 7 4 10

(524) V
3 9 14 8
16 6 1 11
5 15 12 2
10 4 7 13

(525) VII
3 9 16 6
10 8 1 15
7 13 12 2
14 4 5 11

(526) IX
3 9 16 6
14 4 5 11
7 13 12 2
10 8 1 15

(527) IV
3 9 16 6
14 8 1 11
2 12 13 7
15 5 4 10

(528) IV
3 9 16 6
14 8 1 11
5 15 10 4
12 2 7 13

(529) VII
3 9 16 6
15 8 1 10
2 13 12 7
14 4 5 11

(530) I
3 10 5 16
13 8 11 2
12 1 14 7
6 15 4 9

(531) VII
3 10 7 14
13 4 9 8
12 5 16 1
6 15 2 11

(532) I
3 10 8 13
16 5 11 2
9 4 14 7
6 15 1 12

(533) VI
3 10 13 8
12 6 1 15
5 11 16 2
14 7 4 9

(534) VI
3 10 13 8
15 6 1 12
2 11 16 5
14 7 4 9

(535) III
3 10 13 8
16 5 2 11
6 15 12 1
9 4 7 14

(536) I
3 10 15 6
13 8 1 12
2 11 14 7
16 5 4 9

(537) I
3 10 15 6
16 5 4 9
2 11 14 7
13 8 1 12

(538) VII
3 10 15 6
16 7 2 9
1 12 13 8
14 5 4 11

(539) III
3 10 16 5
13 8 2 11
6 15 9 4
12 1 7 14

(540) VI
3 10 16 5
15 6 4 9
2 11 13 8
14 7 1 12

(541) VIII
3 11 14 6
13 10 7 4
2 5 12 15
16 8 1 9

(542) VIII
3 11 14 6
16 10 7 1
2 8 9 15
13 5 4 12

(543) VI
3 12 5 14
13 6 11 4
10 1 16 7
8 15 2 9

(544) X
3 12 5 14
13 15 4 2
8 6 9 11
10 1 6 7

(545) VI
3 12 5 14
15 8 9 2
6 1 16 11
10 13 4 7

(546) X
3 12 5 14
15 13 2 4
6 8 11 9
10 1 16 7

(547) VI
3 12 5 14
16 13 4 1
6 7 10 11
9 2 15 8

(548) IV
3 12 6 13
14 5 11 4
9 2 16 7
8 15 1 10

(549) IV
3 12 13 6
14 5 4 11
2 9 16 7
15 8 1 10

(550) IV
3 12 13 6
14 5 4 11
8 15 10 1
9 2 7 16

(551) IX
3 12 13 6
14 7 2 11
1 10 15 8
16 5 4 9

(552) VII
3 12 13 6
16 5 4 9
1 10 15 8
14 7 2 11

(553) V
3 12 14 5
13 6 4 11
2 9 15 8
16 7 1 10

(554) V
3 12 14 5
13 6 4 11
8 15 9 2
10 1 7 16

(555) IX
3 13 2 16
14 12 7 1
11 5 10 8
6 4 15 9

(556) VI
3 13 4 14
15 8 9 2
6 1 16 11
10 12 5 7

(557) VI
3 13 6 12
15 10 1 8
2 7 16 9
14 4 11 5

(558) III
3 13 6 12
16 2 9 7
10 8 15 1
5 11 4 14

(559) IX
3 13 8 10
14 11 2 7
1 6 15 12
16 4 9 5

(560) I
3 13 8 10
16 2 11 5
9 7 14 4
6 12 1 15

(561) VI	(562) III	(563) IX	(564) VII	(565) I	(566) IV	(567) IX	(568) IX
3 13 10 8	3 13 10 8	3 13 12 6	3 13 12 6	3 13 12 6	3 14 5 12	3 14 7 10	3 14 7 10
15 6 1 12	16 2 5 11	14 7 2 11	15 4 5 10	16 2 7 9	15 2 9 8	16 4 13 1	16 11 6 1
2 11 16 5	6 12 15 1	1 10 15 8	2 9 16 7	5 11 14 4	10 7 16 1	9 5 12 8	2 5 12 15
14 4 7 9	9 7 4 14	16 4 5 9	14 8 1 11	10 8 1 15	6 11 4 13	6 11 2 15	13 4 9 8

(569) IV	(570) IV	(571) IX	(572) IV	(573) VI	(574) VI	(575) V	(576) V
3 14 8 9	3 14 9 8	3 14 11 6	3 14 12 5	3 15 2 14	3 15 2 14	4 1 13 16	4 1 13 16
15 2 12 5	15 2 5 12	16 9 8 1	15 2 8 9	16 6 11 1	16 10 7 1	14 15 3 2	14 15 3 2
10 7 13 4	6 11 16 1	2 7 10 15	6 11 13 4	10 4 13 7	6 4 13 11	7 6 10 11	11 10 6 7
6 11 1 16	10 7 4 13	13 4 5 12	10 7 1 16	5 9 8 12	9 5 12 8	9 12 8 5	5 8 12 9

(577) V	(578) V	(579) VIII	(580) IX	(581) IV	(582) IV	(583) II	(584) II
4 1 13 16	4 1 13 16	4 1 14 15	4 1 14 15	4 1 14 15	4 1 14 15	4 1 14 15	4 1 14 15
15 14 2 3	15 14 2 3	12 9 6 7	13 11 8 2	13 16 3 2	13 16 3 2	16 13 2 3	16 13 2 3
6 7 11 10	10 11 7 6	13 8 11 2	12 6 9 7	7 6 9 12	11 10 5 8	5 8 11 10	9 12 7 6
9 12 8 5	5 8 12 9	5 16 3 10	5 16 3 10	10 11 8 5	6 7 12 9	9 12 7 6	5 8 11 10

(585) VIII	(586) VIII	(587) IX	(588) IX	(589) IV	(590) IV	(591) II	(592) II
4 1 15 14	4 1 15 14	4 1 15 14	4 1 15 14	4 1 15 14	4 1 15 14	4 1 15 14	4 1 15 14
8 11 5 10	12 9 7 6	13 10 8 3	13 12 6 3	13 16 2 3	13 16 2 3	16 13 3 2	16 13 3 2
13 6 12 3	13 8 10 3	12 7 9 6	8 5 11 10	6 7 9 12	10 11 5 8	5 8 10 11	9 12 6 7
9 16 2 7	5 16 2 11	5 16 2 11	9 16 2 7	11 10 8 5	7 6 12 9	9 12 6 7	5 8 10 11

(593) VI	(594) VI	(595) VI	(596) VI	(597) VI	(598) X	(599) VI	(600) VI
4 1 16 13	4 1 16 13	4 1 16 13	4 1 16 13	4 1 16 13	4 1 16 13	4 1 16 13	4 1 16 13
5 14 3 12	5 15 2 12	6 12 5 11	9 14 3 8	9 15 2 8	11 15 6 2	14 15 2 3	14 15 2 3
15 8 9 2	14 8 9 3	15 7 10 2	15 12 5 2	14 12 5 3	14 10 3 7	5 8 9 12	9 12 5 8
10 11 6 7	11 10 7 6	9 14 3 8	6 7 10 11	7 6 11 10	5 8 9 12	11 10 7 6	7 6 11 10

(601) X	(602) VI	(603) VI	(604) VI	(605) VIII	(606) VIII	(607) VIII	(608) VIII
4 1 16 13	4 1 16 13	4 1 16 13	4 1 16 13	4 2 13 15	4 2 13 15	4 2 13 15	4 2 13 15
15 11 2 6	15 12 5 2	15 14 3 2	15 14 3 2	5 14 3 12	9 14 3 8	16 14 3 1	16 14 3 1
10 14 7 3	6 7 10 11	5 8 9 12	9 12 5 8	16 7 10 1	16 11 6 1	5 7 10 12	9 11 6 8
5 8 9 12	9 14 3 8	10 11 6 7	6 7 10 11	9 11 8 6	5 7 12 10	9 11 8 6	5 7 12 10

(609) VI	(610) VI	(611) VI	(612) VI	(613) XII	(614) VI	(615) VI	(616) X
4 2 15 13	4 2 15 13	4 2 15 13	4 2 15 13	4 3 13 14	4 3 14 13	4 3 14 13	4 3 14 13
5 16 1 12	7 16 1 10	14 16 1 3	14 16 1 3	16 15 1 2	5 15 2 12	6 10 7 11	10 16 7 1
14 9 8 3	14 11 6 3	5 9 8 12	7 11 6 10	5 10 8 11	16 10 7 1	15 5 12 2	15 9 2 8
11 7 10 6	9 5 12 8	11 7 10 6	9 5 12 8	9 6 12 7	9 6 11 8	9 16 1 8	5 6 11 12

(617) VI	(618) X	(619) VI	(620) V	(621) I	(622) V	(623) I	(624) VI
4 3 14 13	4 3 14 13	4 3 14 13	4 5 10 15	4 5 10 15	4 5 11 14	4 5 11 14	4 5 12 13
15 10 7 2	16 10 1 7	16 15 2 1	11 14 1 8	14 11 8 1	10 15 1 8	15 10 8 1	7 16 1 10
6 5 12 11	9 15 8 2	5 10 7 12	13 12 7 2	7 2 13 12	13 12 6 3	6 3 13 12	14 11 6 3
9 16 1 8	5 6 11 12	9 6 11 8	6 3 16 9	9 16 3 6	7 2 16 9	9 16 2 7	9 2 15 8

(625) VI	(626) VI	(627) VI	(628) III	(629) VI	(630) VI	(631) VI	(632) III
4 5 12 13	4 5 14 11	4 5 14 11	4 5 14 11	4 5 14 11	4 5 14 11	4 5 14 11	4 5 14 11
14 16 1 3	7 9 2 16	7 16 9 2	9 16 7 2	10 8 1 15	10 15 8 1	15 8 1 10	15 10 1 8
7 11 6 10	10 8 15 1	10 1 8 15	15 10 1 8	7 9 16 2	7 2 9 16	2 9 16 7	9 16 7 2
9 2 15 8	13 12 3 6	13 12 3 6	6 3 12 13	13 12 3 6	13 12 3 6	13 12 3 6	6 3 12 13

(633) VI	(634) VI	(635) III	(636) VI	(637) III	(638) VI	(639) VI	(640) V
4 5 14 11	4 5 15 10	4 5 15 10	4 5 15 10	4 5 15 10	4 5 15 10	4 5 16 9	4 5 16 9
16 9 2 7	6 9 3 16	9 16 6 3	11 14 8 1	14 11 1 8	16 9 3 6	6 15 10 3	10 15 6 3
1 8 15 10	11 8 14 1	14 11 1 8	6 3 9 16	9 16 6 3	1 8 14 11	11 2 7 14	13 12 1 8
13 12 3 6	13 12 2 7	7 2 12 13	13 12 2 7	7 2 12 13	13 12 2 7	13 12 1 8	7 2 11 14

```
(641)    VI      (642)     V      (643)    XI      (644)    IX      (645)    VI      (646)     I      (647)     I      (648)    II
 4  5 16  9       4  5 16  9       4  5 16  9       4  5 16  9       4  5 16  9       4  5 16  9       4  5 16  9       4  6  9 15
11  7  2 14      11 14  7  2      13  8  1 12      13 10  3  8      14  7  2 11      14 11  2  7      15 10  3  6      11 13  2  8
 6 10 15  3      13 12  1  8       3 10 15  6       2  7 14 11       3 10 15  6       1  8 13 12       1  8 13 12      14 12  7  1
13 12  1  8       6  3 10 15      14 11  2  7      15 12  1  6      13 12  1  8      15 10  3  6      14 11  2  7       5  3 16 10

(649)    IV      (650)   XII      (651)    VI      (652)     X      (653)    VI      (654)    VI      (655)     X      (656)    VI
 4  6  9 15       4  6  9 15       4  6 11 13       4  6 11 13       4  6 11 13       4  6 11 13       4  6 11 13       4  6 11 13
13 11  8  2      14 12  1  7       9 15  2  8      10 16  7  1      14 15  2  3      15  9  8  2      16 10  1  7      16 15  2  1
 7  1 14 12      11 13  8  2      14 12  5  3      15  9  2  8       9 12  5  8       5  3 14 12       9 15  8  2       5 10  7 12
10 16  3  5       5  3 16 10       7  1 16 10       5  3 14 12       7  1 16 10      10 16  1  7       5  3 14 12       9  3 14  8

(657)     V      (658)     V      (659)     V      (660)     V      (661)    II      (662)    II      (663)    IV      (664)    IV
 4  6 13 11       4  6 13 11       4  6 13 11       4  6 13 11       4  6 15  9       4  6 15  9       4  6 15  9       4  6 15  9
 9 15  8  2       9 15  8  2      15  9  2  8      15  9  2  8      11 13  8  2      11 13  8  2      13 11  2  8      13 11  2  8
 7  1 10 16      16 10  1  7       1  7 16 10      10 16  7  1       5  3 10 16      14 12  1  7       1  7 14 12      10 16  5  3
14 12  3  5       5  3 12 14      14 12  3  5       5  3 12 14      14 12  1  7       5  3 10 16      16 10  3  5       7  1 12 14

(665)    VI      (666)   XII      (667)    VI      (668)    II      (669)    IV      (670)    VI      (671)    VI      (672)    VI
 4  6 15  9       4  6 15  9       4  6 15  9       4  7  9 14       4  7  9 14       4  7 10 13       4  7 10 13       4  7 10 13
14 12  7  1      14 12  7  1      16 10  5  3      10 13  3  8      13 10  8  3       9 14  3  8      14  9  8  3      14 16  1  3
 3  5 10 16      11 13  2  8       1  7 12 14      15 12  6  1       6  1 15 12      15 12  5  2       5  2 15 12       5  9  8 12
13 11  2  8       5  3 10 16      13 11  2  8       5  2 16 11      11 16  2  5       6  1 16 11      11 16  1  6      11  2 15  6

(673)    VI      (674)     V      (675)     V      (676)     V      (677)     V      (678)    II      (679)    II      (680)    VI
 4  7 10 13       4  7 13 10       4  7 13 10       4  7 13 10       4  7 13 10       4  7 14  9       4  7 14  9       4  7 14  9
15 14  3  2       9 14  8  3       9 14  8  3      14  9  3  8      14  9  3  8      10 13  8  3      10 13  8  3      12  6  1 15
 9 12  5  8       6  1 11 16      16 11  1  6       1  6 16 11      11 16  6  1       5  2 11 16      15 12  1  6       5 11 16  2
 6  1 16 11      15 12  2  5       5  2 12 15      15 12  2  5       5  2 12 15      15 12  1  6       5  2 11 16      13 10  3  8

(681)    IV      (682)    IV      (683)    VI      (684)     X      (685)     X      (686)  VIII      (687)  VIII      (688)  VIII
 4  7 14  9       4  7 14  9       4  7 14  9       4  8  9 13       4  8  9 13       4  8 13  9       4  8 13  9       4  8 13  9
13 10  3  8      13 10  3  8      15  6  1 12      11 15  6  2      15 11  2  6      10 11  6  7      15 11  6  2      15 11  6  2
 1  6 15 12      11 16  5  2       2 11 16  5      14 10  3  7      10 14  7  3      15 14  3  2       1  5 12 16      10 14  3  7
16 11  2  5       6  1 12 15      13 10  3  8       5  1 16 12       5  1 16 12       5  1 12 16      14 10  3  7       5  1 12 16

(689)    XI      (690)     I      (691)     I      (692)   VII      (693)   VII      (694)    VI      (695)   III      (696)    VI
 4  9  6 15       4  9  6 15       4  9  7 14       4  9  8 13       4  9  8 13       4  9 14  7       4  9 14  7       4  9 14  7
11  8 13  2      14  7 12  1      15  6 12  1      10  7 14  3      14  3 10  7      11  5  2 16      15  6  1 12      16  5  2 11
14  1 12  7      11  2 13  8      10  3 13  8      15  2 11  6      11  6 15  2       6 12 15  1       5 16 11  2       1 12 15  6
 5 16  3 10       5 16  3 10       5 16  2 11       5 16  1 12       5 16  1 12      13  8  3 10      10  3  8 13      13  8  3 10

(697)    VI      (698)   III      (699)    VI      (700)   VII      (701)    IX      (702)     I      (703)   VII      (704)     I
 4  9 15  6       4  9 15  6       4  9 15  6       4  9 16  5       4  9 16  5       4  9 16  5       4  9 16  5       4  9 16  5
10  5  3 16      14  7  1 12      16  5  3 10      11  8  1 14      13  6  3 12      14  7  2 11      14  8  1 11      15  6  3 10
 7 12 14  1       5 16 10  3       1 12 14  7       6 15 10  3       2 11 14  7       1 12 13  8       3 15 10  6       1 12 13  8
13  8  2 11      11  2  8 13      13  8  2 11      13  2  7 12      15  8  1 10      15  6  3 10      13  2  7 12      14  7  2 11

(705)    IV      (706)    VI      (707)    VI      (708)    VI      (709)     V      (710)     V      (711)    IX      (712)    IV
 4 10  5 15       4 10  7 13       4 10  7 13       4 10  7 13       4 10 13  7       4 10 13  7       4 10 15  5       4 10 15  5
13  7 12  2      14  8  9  3      14 15  2  3      15  5 12  2      15  5  2 12      15  5  2 12      13  3  6 12      13  7  2 12
11  1 14  8       5  1 16 12       5  8  9 12       9  3 14  8       1 11 16  6       6 16 11  1       8 14 11  1       1 11 14  8
 6 16  3  9      11 15  2  6      11  1 16  6       6 16  1 11      14  8  3  9       9  3  8 14       9  7  2 16      16  6  3  9

(713)    IV      (714)   VII      (715)    IV      (716)    VI      (717)   VII      (718)    VI      (719)    VI      (720)     V
 4 10 15  5       4 10 15  5       4 11  5 14       4 11  6 13       4 11  6 13       4 11  6 13       4 11  6 13       4 11 13  6
13  7  2 12      16  7  2  9      13  6 12  3      14  5 12  3      15  2  9  8      15 14  3  2      16  7 10  1      14  5  3 12
 6 16  9  3       1 14 11  8      10  1 15  8       9  2 15  8      10  7 16  1       5  8  9 12       5  2 15 12       1 10 16  7
11  1  8 14      13  3  6 12       7 16  2  9       7 16  1 10       5 14  3 12      10  1 16  7       9 14  3  8      15  8  2  9
```

(721)   V
4 11 13 6
14 5 3 12
7 16 10 1
9 2 8 15

(722)   IV
4 11 14 5
13 6 3 12
1 10 15 8
16 7 2 9

(723)   IV
4 11 14 5
13 6 3 12
7 16 9 2
10 1 8 15

(724)   XI
4 11 14 5
15 2 7 10
6 13 12 3
9 8 1 16

(725)   VI
4 12 5 13
14 6 11 3
7 1 16 10
9 15 2 8

(726)   VIII
4 12 13 5
14 9 8 3
1 6 11 16
15 7 2 10

(727)   VIII
4 12 13 5
15 9 8 2
1 7 10 16
14 6 3 11

(728)   IX
4 13 2 15
16 6 11 1
9 3 14 8
5 12 7 10

(729)   IX
4 13 2 15
16 10 7 1
5 3 14 12
9 8 11 6

(730)   IV
4 13 6 11
16 1 10 7
9 8 15 2
5 12 3 14

(731)   IV
4 13 7 10
16 1 11 6
9 8 14 3
5 12 2 15

(732)   IX
4 13 8 9
15 3 14 2
10 6 11 7
5 12 1 16

(733)   IX
4 13 8 9
15 12 5 2
1 6 11 16
14 3 10 7

(734)   IV
4 13 10 7
16 1 6 11
5 12 15 2
9 8 3 14

(735)   IV
4 13 11 6
16 1 7 10
5 12 14 3
9 8 2 15

(736)   IX
4 13 12 5
14 11 6 3
1 8 9 16
15 2 7 10

(737)   IX
4 13 12 5
15 10 7 2
1 8 9 16
14 3 6 11

(738)   VII
4 14 3 13
15 2 9 8
10 7 16 1
5 11 6 12

(739)   VI
4 14 3 13
15 12 5 2
6 7 10 11
9 1 16 8

(740)   VI
4 14 3 13
16 7 10 1
5 2 15 12
9 11 6 8

(741)   III
4 14 5 11
15 1 10 8
9 7 16 2
6 12 3 13

(742)   VI
4 14 5 11
15 8 1 10
2 9 16 7
13 3 12 6

(743)   VI
4 14 5 11
16 9 2 7
1 8 15 10
13 3 12 6

(744)   I
4 14 7 9
15 1 12 6
10 8 13 3
5 11 2 16

(745)   VI
4 14 7 9
15 6 1 12
2 11 16 5
13 3 10 8

(746)   III
4 14 9 7
15 1 6 12
5 11 16 2
10 8 3 13

(747)   VI
4 14 9 7
16 5 2 11
1 12 15 6
13 3 8 10

(748)   I
4 14 11 5
15 1 8 10
6 12 13 3
9 7 2 16

(749)   VII
4 14 11 5
16 3 6 9
1 10 15 8
13 7 2 12

(750)   VI
4 15 5 10
16 9 3 6
1 8 14 11
13 2 12 7

(751)   VI
4 15 6 9
16 10 5 3
1 7 12 14
13 2 11 8

(752)   VI
4 15 9 6
16 5 3 10
1 12 14 7
13 2 8 11

(753)   X
5 1 16 12
10 14 7 3
8 4 9 13
11 15 2 6

(754)   X
5 1 16 12
14 10 3 7
4 8 13 9
11 15 2 6

(755)   VI
5 2 15 12
3 13 4 14
16 8 9 1
10 11 6 7

(756)   VI
5 2 15 12
4 11 6 13
16 7 10 1
9 14 3 8

(757)   VII
5 2 15 12
4 13 8 9
14 3 10 7
11 16 1 6

(758)   VII
5 2 15 12
8 9 4 13
10 7 14 3
11 16 1 6

(759)   X
5 2 15 12
8 16 9 1
11 3 6 14
10 13 4 7

(760)   VI
5 2 15 12
10 16 1 7
11 13 4 6
8 3 14 9

(761)   VI
5 2 15 12
11 16 1 6
10 13 4 7
8 3 14 9

(762)   X
5 2 15 12
16 8 1 9
3 11 14 6
10 13 4 7

(763)   VI
5 2 15 12
16 11 6 1
4 7 10 13
9 14 3 8

(764)   VI
5 2 15 12
16 13 4 1
3 8 9 14
10 11 6 7

(765)   VII
5 2 16 11
4 15 1 14
13 10 8 3
12 7 9 6

(766)   IV
5 2 16 11
12 15 1 6
9 14 4 7
8 3 13 10

(767)   VII
5 2 16 11
14 15 1 4
3 10 8 13
12 7 9 6

(768)   II
5 2 16 11
15 12 6 1
4 7 9 14
10 13 3 8

(769)   VI
5 3 14 12
4 10 7 13
16 6 11 1
9 15 2 8

(770)   X
5 3 14 12
8 16 9 1
10 2 7 15
11 13 4 6

(771)   VI
5 3 14 12
10 16 1 7
11 13 4 6
8 2 15 9

(772)   VI
5 3 14 12
11 16 1 6
10 13 4 7
8 2 15 9

(773)   X
5 3 14 12
16 8 1 9
2 10 15 7
11 13 4 6

(774)   VI
5 3 14 12
16 10 7 1
4 6 11 13
9 15 2 8

(775)   VII
5 3 16 10
4 14 1 15
13 11 8 2
12 6 9 7

(776)   VI
5 3 16 10
6 9 4 15
11 8 13 2
12 14 1 7

(777)   VI
5 3 16 10
11 13 8 2
6 4 9 15
12 14 1 7

(778)   IV
5 3 16 10
12 14 1 7
9 15 4 6
8 2 13 11

(779)   II
5 3 16 10
14 12 7 1
4 6 9 15
11 13 2 8

(780)   VI
5 3 16 10
15 9 4 6
2 8 13 11
12 14 1 7

(781)   VII
5 3 16 10
15 14 1 4
2 11 8 13
12 6 9 7

(782)   VII
5 4 13 12
10 7 2 15
8 9 16 1
11 14 3 6

(783)   VII
5 4 13 12
11 6 3 14
8 9 16 1
10 15 2 7

(784)   V
5 4 14 11
9 16 2 7
12 13 3 6
8 1 15 10

(785)   I
5 4 14 11
16 9 7 2
3 6 12 13
10 15 1 8

(786)   VII
5 4 14 11
16 13 3 2
1 8 10 15
12 9 7 6

(787)   V
5 4 15 10
9 16 3 6
12 13 2 7
8 1 14 11

(788)   I
5 4 15 10
16 9 6 3
2 7 12 13
11 14 1 8

(789)   III
5 4 16 9
10 15 3 6
11 14 2 7
8 1 13 12

(790)   III
5 4 16 9
11 14 2 7
10 15 3 6
8 1 13 12

(791)   VI
5 4 16 9
14 11 7 2
10 15 3 6
12 13 1 8

(792)   VI
5 4 16 9
15 10 6 3
2 7 11 14
12 13 1 8

(793)   VI
5 6 11 12
16 9 8 1
3 4 13 14
10 15 2 7

(794)   X
5 8 10 11
13 16 2 3
4 9 7 14
12 1 15 6

(795)   VII
5 8 10 11
16 9 7 2
1 4 14 15
12 13 3 6

(796)   X
5 8 11 10
13 16 3 2
4 9 6 15
12 1 14 7

(797)   V
5 9 12 8
16 4 1 13
2 14 15 3
11 7 6 10

(798)   V
5 9 12 8
16 4 1 13
3 15 14 2
10 6 7 11

(799)   VII
5 10 8 11
14 7 9 4
3 2 16 13
12 15 1 6

(800)   X
5 10 8 11
15 14 4 1
2 7 9 16
12 3 13 6

(801) VI	(802) VI	(803) III	(804) VI	(805) VII	(806) VI	(807) VI	(808) III
5 10 11 8	5 10 11 8	5 10 11 8	5 11 6 12	5 11 8 10	5 11 10 8	5 11 10 8	5 11 10 8
14 4 1 15	15 4 1 14	16 3 2 13	16 13 4 1	15 6 9 4	14 4 1 15	15 4 1 14	16 2 3 13
3 13 16 2	2 13 16 3	4 15 14 1	3 8 9 14	2 3 16 13	3 13 16 2	2 13 16 3	4 14 15 1
12 7 6 9	12 7 6 9	9 6 7 12	10 2 15 7	12 14 1 7	12 6 7 9	12 6 7 9	9 7 6 12

(809) IV	(810) IV	(811) X	(812) X	(813) VI	(814) X	(815) VI	(816) VI
5 12 9 8	5 12 9 8	5 13 4 12	5 13 4 12	5 14 3 12	5 14 4 11	5 15 2 12	5 15 2 12
14 3 2 15	15 2 3 14	16 8 1 9	16 8 1 9	16 11 6 1	15 10 8 1	16 9 8 1	16 10 7 1
4 13 16 1	4 13 16 1	2 10 15 7	3 11 14 6	4 7 10 13	2 3 13 16	3 4 13 14	4 6 11 13
11 6 7 10	10 7 6 11	11 3 14 6	10 2 15 7	9 2 15 8	12 7 9 6	10 6 11 7	9 3 14 8

(817) IV	(818) II	(819) VI	(820) X	(821) VI	(822) VI	(823) X	(824) VI
6 1 15 12	6 1 15 12	6 1 16 11	6 1 16 11	6 1 16 11	6 1 16 11	6 1 16 11	6 1 16 11
11 16 2 5	16 11 5 2	3 12 5 14	7 15 10 2	9 15 2 8	12 15 2 5	15 7 2 10	15 12 5 2
10 13 3 8	3 8 10 13	15 8 9 2	12 4 5 13	12 14 3 5	9 14 3 8	4 12 13 5	3 8 9 14
7 4 14 9	9 14 4 7	10 13 4 7	9 14 3 8	7 4 13 10	7 4 13 10	9 14 3 8	10 13 4 7

(825) VI	(826) VI	(827) V	(828) I	(829) VII	(830) VI	(831) VII	(832) VI
6 2 15 11	6 2 15 11	6 3 13 12	6 3 13 12	6 3 14 11	6 3 14 11	6 3 14 11	6 3 14 11
7 16 1 10	12 16 1 5	10 15 1 8	15 10 8 1	4 13 12 5	7 16 1 10	12 5 4 13	12 16 1 5
12 13 4 5	7 13 4 10	11 14 4 5	4 5 11 14	15 2 7 10	12 13 4 5	7 10 15 2	7 13 4 10
9 3 14 8	9 3 14 8	7 2 16 9	9 16 2 7	9 16 1 8	9 2 15 8	9 16 1 8	9 2 15 8

(833) VI	(834) III	(835) III	(836) VI	(837) VI	(838) V	(839) I	(840) VI
6 3 15 10	6 3 15 10	6 3 15 10	6 3 15 10	6 3 15 10	6 3 16 9	6 3 16 9	6 4 13 11
4 9 5 16	9 16 4 5	12 13 1 8	13 12 8 1	16 9 5 4	10 15 4 5	15 10 5 4	9 15 2 8
13 8 12 1	12 13 1 8	9 16 4 5	4 5 9 16	1 8 12 13	11 14 1 8	1 8 11 14	12 14 3 5
11 14 2 7	7 2 14 11	7 2 14 11	11 14 2 7	11 14 2 7	7 2 13 12	12 13 2 7	7 1 16 10

(841) VI	(842) VI	(843) IV	(844) II	(845) VII	(846) VI	(847) X	(848) VI
6 4 13 11	6 4 13 11	6 4 15 9	6 4 15 9	6 4 15 9	6 7 12 9	6 7 12 9	6 7 12 9
12 15 2 5	15 9 8 2	11 13 2 8	13 11 8 2	16 13 2 3	14 4 1 15	14 15 4 1	15 4 1 14
9 14 3 8	3 5 12 14	10 16 3 5	3 5 10 16	1 12 7 14	3 13 16 2	3 10 5 16	2 13 16 3
7 1 16 10	10 16 1 7	7 1 14 12	12 14 1 7	11 5 10 8	11 10 5 8	11 2 13 8	11 10 5 8

(849) VI	(850) III	(851) VI	(852) V	(853) V	(854) IV	(855) IV	(856) VI
6 9 12 7	6 9 12 7	6 9 12 7	6 10 11 7	6 10 11 7	6 11 10 7	6 11 10 7	6 12 7 9
13 3 2 16	15 4 1 14	16 3 2 13	15 3 2 14	15 3 2 14	13 4 1 16	16 1 4 13	14 4 1 15
4 14 15 1	3 16 13 2	1 14 15 4	1 13 16 4	4 16 13 1	3 14 15 2	3 14 15 2	3 13 16 2
11 8 5 10	10 5 8 11	11 8 5 10	12 8 5 9	9 5 8 12	12 5 8 9	9 8 5 12	11 5 10 8

(857) VI	(858) VII	(859) VI	(860) III	(861) VI	(862) VI	(863) X	(864) VI
6 12 7 9	6 12 7 9	6 12 9 7	6 12 9 7	6 12 9 7	6 13 4 11	6 14 3 11	6 15 3 10
15 4 1 14	16 5 10 3	13 3 2 16	15 1 4 14	16 3 2 13	15 12 5 2	15 7 2 10	16 9 5 4
2 13 16 3	1 4 15 14	4 14 15 1	3 13 16 2	1 14 15 4	3 8 9 14	4 12 13 5	1 8 12 13
11 5 10 8	11 13 2 8	11 5 8 10	10 8 5 11	11 5 8 10	10 1 16 7	9 1 16 8	11 2 14 7

(865) X	(866) X	(867) VII	(868) VII	(869) VII	(870) VII	(871) X	(872) X
7 1 16 10	7 1 16 10	7 2 15 10	7 2 15 10	7 2 16 9	7 2 16 9	7 4 14 9	7 4 14 9
6 14 11 3	14 6 3 11	4 13 12 5	12 5 4 13	6 15 1 12	12 15 1 6	5 16 2 11	11 16 2 5
12 4 5 13	4 12 13 5	14 3 6 11	6 11 14 3	11 14 4 5	5 14 4 11	12 13 3 6	6 13 3 12
9 15 2 8	9 15 2 8	9 16 1 8	9 16 1 8	10 3 13 8	10 3 13 8	10 1 15 8	10 1 15 8

(873) VII	(874) X	(875) VII	(876) X	(877) X	(878) VII	(879) VII	(880) X
7 4 14 9	7 5 12 10	7 6 11 10	7 6 12 9	7 11 6 10	7 12 5 10	7 12 6 9	7 14 4 9
16 13 3 2	16 4 1 13	14 3 2 15	15 14 4 1	16 4 1 13	14 3 2 15	16 5 11 2	15 6 12 1
1 12 6 15	2 14 15 3	4 13 16 1	2 11 5 16	2 14 15 3	4 13 16 1	1 4 14 15	2 3 13 16
10 5 11 8	9 11 6 8	9 12 5 8	10 3 13 8	9 5 12 8	9 6 11 8	10 13 3 8	10 11 5 8

A CATALOGUE OF SELECTED DOVER BOOKS
IN ALL FIELDS OF INTEREST

A CATALOGUE OF SELECTED DOVER BOOKS
IN ALL FIELDS OF INTEREST

AMERICA'S OLD MASTERS, James T. Flexner. Four men emerged unexpectedly from provincial 18th century America to leadership in European art: Benjamin West, J. S. Copley, C. R. Peale, Gilbert Stuart. Brilliant coverage of lives and contributions. Revised, 1967 edition. 69 plates. 365pp. of text.
21806-6 Paperbound $3.00

FIRST FLOWERS OF OUR WILDERNESS: AMERICAN PAINTING, THE COLONIAL PERIOD, James T. Flexner. Painters, and regional painting traditions from earliest Colonial times up to the emergence of Copley, West and Peale Sr., Foster, Gustavus Hesselius, Feke, John Smibert and many anonymous painters in the primitive manner. Engaging presentation, with 162 illustrations. xxii + 368pp.
22180-6 Paperbound $3.50

THE LIGHT OF DISTANT SKIES: AMERICAN PAINTING, 1760-1835, James T. Flexner. The great generation of early American painters goes to Europe to learn and to teach: West, Copley, Gilbert Stuart and others. Allston, Trumbull, Morse; also contemporary American painters—primitives, derivatives, academics—who remained in America. 102 illustrations. xiii + 306pp.
22179-2 Paperbound $3.50

A HISTORY OF THE RISE AND PROGRESS OF THE ARTS OF DESIGN IN THE UNITED STATES, William Dunlap. Much the richest mine of information on early American painters, sculptors, architects, engravers, miniaturists, etc. The only source of information for scores of artists, the major primary source for many others. Unabridged reprint of rare original 1834 edition, with new introduction by James T. Flexner, and 394 new illustrations. Edited by Rita Weiss. 6⅜ x 9⅝.
21695-0, 21696-9, 21697-7 Three volumes, Paperbound $15.00

EPOCHS OF CHINESE AND JAPANESE ART, Ernest F. Fenollosa. From primitive Chinese art to the 20th century, thorough history, explanation of every important art period and form, including Japanese woodcuts; main stress on China and Japan, but Tibet, Korea also included. Still unexcelled for its detailed, rich coverage of cultural background, aesthetic elements, diffusion studies, particularly of the historical period. 2nd, 1913 edition. 242 illustrations. lii + 439pp. of text.
20364-6, 20365-4 Two volumes, Paperbound $6.00

THE GENTLE ART OF MAKING ENEMIES, James A. M. Whistler. Greatest wit of his day deflates Oscar Wilde, Ruskin, Swinburne; strikes back at inane critics, exhibitions, art journalism; aesthetics of impressionist revolution in most striking form. Highly readable classic by great painter. Reproduction of edition designed by Whistler. Introduction by Alfred Werner. xxxvi + 334pp.
21875-9 Paperbound $3.00

JOHANN SEBASTIAN BACH, Philipp Spitta. One of the great classics of musicology, this definitive analysis of Bach's music (and life) has never been surpassed. Lucid, nontechnical analyses of hundreds of pieces (30 pages devoted to St. Matthew Passion, 26 to B Minor Mass). Also includes major analysis of 18th-century music. 450 musical examples. 40-page musical supplement. Total of xx + 1799pp.

(EUK) 22278-0, 22279-9 Two volumes, Clothbound $25.00

MOZART AND HIS PIANO CONCERTOS, Cuthbert Girdlestone. The only full-length study of an important area of Mozart's creativity. Provides detailed analyses of all 23 concertos, traces inspirational sources. 417 musical examples. Second edition. 509pp.

21271-8 Paperbound $4.50

THE PERFECT WAGNERITE: A COMMENTARY ON THE NIBLUNG'S RING, George Bernard Shaw. Brilliant and still relevant criticism in remarkable essays on Wagner's Ring cycle, Shaw's ideas on political and social ideology behind the plots, role of Leitmotifs, vocal requisites, etc. Prefaces. xxi + 136pp.

(USO) 21707-8 Paperbound $1.75

DON GIOVANNI, W. A. Mozart. Complete libretto, modern English translation; biographies of composer and librettist; accounts of early performances and critical reaction. Lavishly illustrated. All the material you need to understand and appreciate this great work. Dover Opera Guide and Libretto Series; translated and introduced by Ellen Bleiler. 92 illustrations. 209pp.

21134-7 Paperbound $2.00

BASIC ELECTRICITY, U. S. Bureau of Naval Personel. Originally a training course, best non-technical coverage of basic theory of electricity and its applications. Fundamental concepts, batteries, circuits, conductors and wiring techniques, AC and DC, inductance and capacitance, generators, motors, transformers, magnetic amplifiers, synchros, servomechanisms, etc. Also covers blue-prints, electrical diagrams, etc. Many questions, with answers. 349 illustrations. x + 448pp. 6½ x 9¼.

20973-3 Paperbound $3.50

REPRODUCTION OF SOUND, Edgar Villchur. Thorough coverage for laymen of high fidelity systems, reproducing systems in general, needles, amplifiers, preamps, loudspeakers, feedback, explaining physical background. "A rare talent for making technicalities vividly comprehensible," R. Darrell, *High Fidelity*. 69 figures. iv + 92pp.

21515-6 Paperbound $1.35

HEAR ME TALKIN' TO YA: THE STORY OF JAZZ AS TOLD BY THE MEN WHO MADE IT, Nat Shapiro and Nat Hentoff. Louis Armstrong, Fats Waller, Jo Jones, Clarence Williams, Billy Holiday, Duke Ellington, Jelly Roll Morton and dozens of other jazz greats tell how it was in Chicago's South Side, New Orleans, depression Harlem and the modern West Coast as jazz was born and grew. xvi + 429pp.

21726-4 Paperbound $3.95

FABLES OF AESOP, translated by Sir Roger L'Estrange. A reproduction of the very rare 1931 Paris edition; a selection of the most interesting fables, together with 50 imaginative drawings by Alexander Calder. v + 128pp. 6½x9¼.

21780-9 Paperbound $1.50

AGAINST THE GRAIN (A REBOURS), Joris K. Huysmans. Filled with weird images, evidences of a bizarre imagination, exotic experiments with hallucinatory drugs, rich tastes and smells and the diversions of its sybarite hero Duc Jean des Esseintes, this classic novel pushed 19th-century literary decadence to its limits. Full unabridged edition. Do not confuse this with abridged editions generally sold. Introduction by Havelock Ellis. xlix + 206pp. 22190-3 Paperbound $2.50

VARIORUM SHAKESPEARE: HAMLET. Edited by Horace H. Furness; a landmark of American scholarship. Exhaustive footnotes and appendices treat all doubtful words and phrases, as well as suggested critical emendations throughout the play's history. First volume contains editor's own text, collated with all Quartos and Folios. Second volume contains full first Quarto, translations of Shakespeare's sources (Belleforest, and Saxo Grammaticus), Der Bestrafte Brudermord, and many essays on critical and historical points of interest by major authorities of past and present. Includes details of staging and costuming over the years. By far the best edition available for serious students of Shakespeare. Total of xx + 905pp. 21004-9, 21005-7, 2 volumes, Paperbound $7.00

A LIFE OF WILLIAM SHAKESPEARE, Sir Sidney Lee. This is the standard life of Shakespeare, summarizing everything known about Shakespeare and his plays. Incredibly rich in material, broad in coverage, clear and judicious, it has served thousands as the best introduction to Shakespeare. 1931 edition. 9 plates. xxix + 792pp. 21967-4 Paperbound $4.50

MASTERS OF THE DRAMA, John Gassner. Most comprehensive history of the drama in print, covering every tradition from Greeks to modern Europe and America, including India, Far East, etc. Covers more than 800 dramatists, 2000 plays, with biographical material, plot summaries, theatre history, criticism, etc. "Best of its kind in English," New Republic. 77 illustrations. xxii + 890pp. 20100-7 Clothbound $10.00

THE EVOLUTION OF THE ENGLISH LANGUAGE, George McKnight. The growth of English, from the 14th century to the present. Unusual, non-technical account presents basic information in very interesting form: sound shifts, change in grammar and syntax, vocabulary growth, similar topics. Abundantly illustrated with quotations. Formerly Modern English in the Making. xii + 590pp. 21932-1 Paperbound $3.50

AN ETYMOLOGICAL DICTIONARY OF MODERN ENGLISH, Ernest Weekley. Fullest, richest work of its sort, by foremost British lexicographer. Detailed word histories, including many colloquial and archaic words; extensive quotations. Do not confuse this with the Concise Etymological Dictionary, which is much abridged. Total of xxvii + 830pp. 6½ x 9¼. 21873-2, 21874-0 Two volumes, Paperbound $7.90

FLATLAND: A ROMANCE OF MANY DIMENSIONS, E. A. Abbott. Classic of science-fiction explores ramifications of life in a two-dimensional world, and what happens when a three-dimensional being intrudes. Amusing reading, but also useful as introduction to thought about hyperspace. Introduction by Banesh Hoffmann. 16 illustrations. xx + 103pp. 20001-9 Paperbound $1.00

POEMS OF ANNE BRADSTREET, edited with an introduction by Robert Hutchinson. A new selection of poems by America's first poet and perhaps the first significant woman poet in the English language. 48 poems display her development in works of considerable variety—love poems, domestic poems, religious meditations, formal elegies, "quaternions," etc. Notes, bibliography. viii + 222pp.

22160-1 Paperbound $2.50

THREE GOTHIC NOVELS: THE CASTLE OF OTRANTO BY HORACE WALPOLE; VATHEK BY WILLIAM BECKFORD; THE VAMPYRE BY JOHN POLIDORI, WITH FRAGMENT OF A NOVEL BY LORD BYRON, edited by E. F. Bleiler. The first Gothic novel, by Walpole; the finest Oriental tale in English, by Beckford; powerful Romantic supernatural story in versions by Polidori and Byron. All extremely important in history of literature; all still exciting, packed with supernatural thrills, ghosts, haunted castles, magic, etc. xl + 291pp.

21232-7 Paperbound $3.00

THE BEST TALES OF HOFFMANN, E. T. A. Hoffmann. 10 of Hoffmann's most important stories, in modern re-editings of standard translations: Nutcracker and the King of Mice, Signor Formica, Automata, The Sandman, Rath Krespel, The Golden Flowerpot, Master Martin the Cooper, The Mines of Falun, The King's Betrothed, A New Year's Eve Adventure. 7 illustrations by Hoffmann. Edited by E. F. Bleiler. xxxix + 419pp. 21793-0 Paperbound $3.00

GHOST AND HORROR STORIES OF AMBROSE BIERCE, Ambrose Bierce. 23 strikingly modern stories of the horrors latent in the human mind: The Eyes of the Panther, The Damned Thing, An Occurrence at Owl Creek Bridge, An Inhabitant of Carcosa, etc., plus the dream-essay, Visions of the Night. Edited by E. F. Bleiler. xxii + 199pp. 20767-6 Paperbound $2.00

BEST GHOST STORIES OF J. S. LEFANU, J. Sheridan LeFanu. Finest stories by Victorian master often considered greatest supernatural writer of all. Carmilla, Green Tea, The Haunted Baronet, The Familiar, and 12 others. Most never before available in the U. S. A. Edited by E. F. Bleiler. 8 illustrations from Victorian publications. xvii + 467pp. 20415-4 Paperbound $3.00

MATHEMATICAL FOUNDATIONS OF INFORMATION THEORY, A. I. Khinchin. Comprehensive introduction to work of Shannon, McMillan, Feinstein and Khinchin, placing these investigations on a rigorous mathematical basis. Covers entropy concept in probability theory, uniqueness theorem, Shannon's inequality, ergodic sources, the E property, martingale concept, noise, Feinstein's fundamental lemma, Shanon's first and second theorems. Translated by R. A. Silverman and M. D. Friedman. iii + 120pp. 60434-9 Paperbound $2.00

SEVEN SCIENCE FICTION NOVELS, H. G. Wells. The standard collection of the great novels. Complete, unabridged. *First Men in the Moon, Island of Dr. Moreau, War of the Worlds, Food of the Gods, Invisible Man, Time Machine, In the Days of the Comet.* Not only science fiction fans, but every educated person owes it to himself to read these novels. 1015pp. (USO) 20264-X Clothbound $6.00

LAST AND FIRST MEN AND STAR MAKER, TWO SCIENCE FICTION NOVELS, Olaf Stapledon. Greatest future histories in science fiction. In the first, human intelligence is the "hero," through strange paths of evolution, interplanetary invasions, incredible technologies, near extinctions and reemergences. Star Maker describes the quest of a band of star rovers for intelligence itself, through time and space: weird inhuman civilizations, crustacean minds, symbiotic worlds, etc. Complete, unabridged. v + 438pp. (USO) 21962-3 Paperbound $3.00

THREE PROPHETIC NOVELS, H. G. WELLS. Stages of a consistently planned future for mankind. *When the Sleeper Wakes,* and *A Story of the Days to Come,* anticipate *Brave New World* and *1984,* in the 21st Century; *The Time Machine,* only complete version in print, shows farther future and the end of mankind. All show Wells's greatest gifts as storyteller and novelist. Edited by E. F. Bleiler. x + 335pp. (USO) 20605-X Paperbound $3.00

THE DEVIL'S DICTIONARY, Ambrose Bierce. America's own Oscar Wilde—Ambrose Bierce—offers his barbed iconoclastic wisdom in over 1,000 definitions hailed by H. L. Mencken as "some of the most gorgeous witticisms in the English language." 145pp. 20487-1 Paperbound $1.50

MAX AND MORITZ, Wilhelm Busch. Great children's classic, father of comic strip, of two bad boys, Max and Moritz. Also Ker and Plunk (Plisch und Plumm), Cat and Mouse, Deceitful Henry, Ice-Peter, The Boy and the Pipe, and five other pieces. Original German, with English translation. Edited by H. Arthur Klein; translations by various hands and H. Arthur Klein. vi + 216pp. 20181-3 Paperbound $2.00

PIGS IS PIGS AND OTHER FAVORITES, Ellis Parker Butler. The title story is one of the best humor short stories, as Mike Flannery obfuscates biology and English. Also included, That Pup of Murchison's, The Great American Pie Company, and Perkins of Portland. 14 illustrations. v + 109pp. 21532-6 Paperbound $1.50

THE PETERKIN PAPERS, Lucretia P. Hale. It takes genius to be as stupidly mad as the Peterkins, as they decide to become wise, celebrate the "Fourth," keep a cow, and otherwise strain the resources of the Lady from Philadelphia. Basic book of American humor. 153 illustrations. 219pp. 20794-3 Paperbound $2.00

PERRAULT'S FAIRY TALES, translated by A. E. Johnson and S. R. Littlewood, with 34 full-page illustrations by Gustave Doré. All the original Perrault stories—Cinderella, Sleeping Beauty, Bluebeard, Little Red Riding Hood, Puss in Boots, Tom Thumb, etc.—with their witty verse morals and the magnificent illustrations of Doré. One of the five or six great books of European fairy tales. viii + 117pp. 8⅛ x 11. 22311-6 Paperbound $2.00

OLD HUNGARIAN FAIRY TALES, Baroness Orczy. Favorites translated and adapted by author of the *Scarlet Pimpernel.* Eight fairy tales include "The Suitors of Princess Fire-Fly," "The Twin Hunchbacks," "Mr. Cuttlefish's Love Story," and "The Enchanted Cat." This little volume of magic and adventure will captivate children as it has for generations. 90 drawings by Montagu Barstow. 96pp. (USO) 22293-4 Paperbound $1.95

THE RED FAIRY BOOK, Andrew Lang. Lang's color fairy books have long been children's favorites. This volume includes Rapunzel, Jack and the Bean-stalk and 35 other stories, familiar and unfamiliar. 4 plates, 93 illustrations x + 367pp.
21673-X Paperbound $2.50

THE BLUE FAIRY BOOK, Andrew Lang. Lang's tales come from all countries and all times. Here are 37 tales from Grimm, the Arabian Nights, Greek Mythology, and other fascinating sources. 8 plates, 130 illustrations. xi + 390pp.
21437-0 Paperbound $2.75

HOUSEHOLD STORIES BY THE BROTHERS GRIMM. Classic English-language edition of the well-known tales — Rumpelstiltskin, Snow White, Hansel and Gretel, The Twelve Brothers, Faithful John, Rapunzel, Tom Thumb (52 stories in all). Translated into simple, straightforward English by Lucy Crane. Ornamented with head-pieces, vignettes, elaborate decorative initials and a dozen full-page illustrations by Walter Crane. x + 269pp.
21080-4 Paperbound **$2.00**

THE MERRY ADVENTURES OF ROBIN HOOD, Howard Pyle. The finest modern versions of the traditional ballads and tales about the great English outlaw. Howard Pyle's complete prose version, with every word, every illustration of the first edition. Do not confuse this facsimile of the original (1883) with modern editions that change text or illustrations. 23 plates plus many page decorations. xxii + 296pp.
22043-5 Paperbound $2.75

THE STORY OF KING ARTHUR AND HIS KNIGHTS, Howard Pyle. The finest children's version of the life of King Arthur; brilliantly retold by Pyle, with 48 of his most imaginative illustrations. xviii + 313pp. 6⅛ x 9¼.
21445-1 Paperbound $2.50

THE WONDERFUL WIZARD OF OZ, L. Frank Baum. America's finest children's book in facsimile of first edition with all Denslow illustrations in full color. The edition a child should have. Introduction by Martin Gardner. 23 color plates, scores of drawings. iv + 267pp.
20691-2 Paperbound $3.50

THE MARVELOUS LAND OF OZ, L. Frank Baum. The second Oz book, every bit as imaginative as the Wizard. The hero is a boy named Tip, but the Scarecrow and the Tin Woodman are back, as is the Oz magic. 16 color plates, 120 drawings by John R. Neill. 287pp.
20692-0 Paperbound $2.50

THE MAGICAL MONARCH OF MO, L. Frank Baum. Remarkable adventures in a land even stranger than Oz. The best of Baum's books not in the Oz series. 15 color plates and dozens of drawings by Frank Verbeck. xviii + 237pp.
21892-9 Paperbound $2.25

THE BAD CHILD'S BOOK OF BEASTS, MORE BEASTS FOR WORSE CHILDREN, A MORAL ALPHABET, Hilaire Belloc. Three complete humor classics in one volume. Be kind to the frog, and do not call him names . . . and 28 other whimsical animals. Familiar favorites and some not so well known. Illustrated by Basil Blackwell. 156pp.
(USO) 20749-8 Paperbound $1.50

EAST O' THE SUN AND WEST O' THE MOON, George W. Dasent. Considered the best of all translations of these Norwegian folk tales, this collection has been enjoyed by generations of children (and folklorists too). Includes True and Untrue, Why the Sea is Salt, East O' the Sun and West O' the Moon, Why the Bear is Stumpy-Tailed, Boots and the Troll, The Cock and the Hen, Rich Peter the Pedlar, and 52 more. The only edition with all 59 tales. 77 illustrations by Erik Werenskiold and Theodor Kittelsen. xv + 418pp.　　　　　　　　　　　22521-6 Paperbound $3.50

GOOPS AND HOW TO BE THEM, Gelett Burgess. Classic of tongue-in-cheek humor, masquerading as etiquette book. 87 verses, twice as many cartoons, show mischievous Goops as they demonstrate to children virtues of table manners, neatness, courtesy, etc. Favorite for generations. viii + 88pp. 6½ x 9¼.
　　　　　　　　　　　　　　　　　　22233-0 Paperbound $1.50

ALICE'S ADVENTURES UNDER GROUND, Lewis Carroll. The first version, quite different from the final *Alice in Wonderland,* printed out by Carroll himself with his own illustrations. Complete facsimile of the "million dollar" manuscript Carroll gave to Alice Liddell in 1864. Introduction by Martin Gardner. viii + 96pp. Title and dedication pages in color.　　　　　　　　21482-6 Paperbound $1.25

THE BROWNIES, THEIR BOOK, Palmer Cox. Small as mice, cunning as foxes, exuberant and full of mischief, the Brownies go to the zoo, toy shop, seashore, circus, etc., in 24 verse adventures and 266 illustrations. Long a favorite, since their first appearance in St. Nicholas Magazine. xi + 144pp. 6⅝ x 9¼.
　　　　　　　　　　　　　　　　　　21265-3 Paperbound $1.75

SONGS OF CHILDHOOD, Walter De La Mare. Published (under the pseudonym Walter Ramal) when De La Mare was only 29, this charming collection has long been a favorite children's book. A facsimile of the first edition in paper, the 47 poems capture the simplicity of the nursery rhyme and the ballad, including such lyrics as I Met Eve, Tartary, The Silver Penny. vii + 106pp. (USO) 21972-0 Paperbound
　　　　　　　　　　　　　　　　　　　　　　　　　　$1.25

THE COMPLETE NONSENSE OF EDWARD LEAR, Edward Lear. The finest 19th-century humorist-cartoonist in full: all nonsense limericks, zany alphabets, Owl and Pussycat, songs, nonsense botany, and more than 500 illustrations by Lear himself. Edited by Holbrook Jackson. xxix + 287pp.　　　(USO) 20167-8 Paperbound $2.00

BILLY WHISKERS: THE AUTOBIOGRAPHY OF A GOAT, Frances Trego Montgomery. A favorite of children since the early 20th century, here are the escapades of that rambunctious, irresistible and mischievous goat—Billy Whiskers. Much in the spirit of *Peck's Bad Boy,* this is a book that children never tire of reading or hearing. All the original familiar illustrations by W. H. Fry are included: 6 color plates, 18 black and white drawings. 159pp.　　　　　22345-0 Paperbound $2.00

MOTHER GOOSE MELODIES. Faithful republication of the fabulously rare Munroe and Francis "copyright 1833" Boston edition—the most important Mother Goose collection, usually referred to as the "original." Familiar rhymes plus many rare ones, with wonderful old woodcut illustrations. Edited by E. F. Bleiler. 128pp. 4½ x 6⅜.　　　　　　　　　　　　　　　22577-1 Paperbound $1.00

TWO LITTLE SAVAGES; BEING THE ADVENTURES OF TWO BOYS WHO LIVED AS INDIANS AND WHAT THEY LEARNED, Ernest Thompson Seton. Great classic of nature and boyhood provides a vast range of woodlore in most palatable form, a genuinely entertaining story. Two farm boys build a teepee in woods and live in it for a month, working out Indian solutions to living problems, star lore, birds and animals, plants, etc. 293 illustrations. vii + 286pp.

20985-7 Paperbound $2.50

PETER PIPER'S PRACTICAL PRINCIPLES OF PLAIN & PERFECT PRONUNCIATION. Alliterative jingles and tongue-twisters of surprising charm, that made their first appearance in America about 1830. Republished in full with the spirited woodcut illustrations from this earliest American edition. 32pp. 4½ x 6⅜.

22560-7 Paperbound $1.00

SCIENCE EXPERIMENTS AND AMUSEMENTS FOR CHILDREN, Charles Vivian. 73 easy experiments, requiring only materials found at home or easily available, such as candles, coins, steel wool, etc.; illustrate basic phenomena like vacuum, simple chemical reaction, etc. All safe. Modern, well-planned. Formerly *Science Games for Children*. 102 photos, numerous drawings. 96pp. 6⅛ x 9¼.

21856-2 Paperbound $1.25

AN INTRODUCTION TO CHESS MOVES AND TACTICS SIMPLY EXPLAINED, Leonard Barden. Informal intermediate introduction, quite strong in explaining reasons for moves. Covers basic material, tactics, important openings, traps, positional play in middle game, end game. Attempts to isolate patterns and recurrent configurations. Formerly *Chess*. 58 figures. 102pp. (USO) 21210-6 Paperbound $1.25

LASKER'S MANUAL OF CHESS, Dr. Emanuel Lasker. Lasker was not only one of the five great World Champions, he was also one of the ablest expositors, theorists, and analysts. In many ways, his Manual, permeated with his philosophy of battle, filled with keen insights, is one of the greatest works ever written on chess. Filled with analyzed games by the great players. A single-volume library that will profit almost any chess player, beginner or master. 308 diagrams. xli x 349pp.

20640-8 Paperbound $2.75

THE MASTER BOOK OF MATHEMATICAL RECREATIONS, Fred Schuh. In opinion of many the finest work ever prepared on mathematical puzzles, stunts, recreations; exhaustively thorough explanations of mathematics involved, analysis of effects, citation of puzzles and games. Mathematics involved is elementary. Translated by F. Göbel 194 figures. xxiv + 430pp. 22134-2 Paperbound $4.00

MATHEMATICS, MAGIC AND MYSTERY, Martin Gardner. Puzzle editor for Scientific American explains mathematics behind various mystifying tricks: card tricks, stage "mind reading," coin and match tricks, counting out games, geometric dissections, etc. Probability sets, theory of numbers clearly explained. Also provides more than 400 tricks, guaranteed to work, that you can do. 135 illustrations. xii + 176pp.

20335-2 Paperbound $2.00

"ESSENTIAL GRAMMAR" SERIES

All you really need to know about modern, colloquial grammar. Many educational shortcuts help you learn faster, understand better. Detailed cognate lists teach you to recognize similarities between English and foreign words and roots—make learning vocabulary easy and interesting. Excellent for independent study or as a supplement to record courses.

ESSENTIAL FRENCH GRAMMAR, Seymour Resnick. 2500-item cognate list. 159pp.
(EBE) 20419-7 Paperbound $1.50

ESSENTIAL GERMAN GRAMMAR, Guy Stern and Everett F. Bleiler. Unusual shortcuts on noun declension, word order, compound verbs. 124pp.
(EBE) 20422-7 Paperbound $1.25

ESSENTIAL ITALIAN GRAMMAR, Olga Ragusa. 111pp.
(EBE) 20779-X Paperbound $1.25

ESSENTIAL JAPANESE GRAMMAR, Everett F. Bleiler. In Romaji transcription; no characters needed. Japanese grammar is regular and simple. 156pp.
21027-8 Paperbound $1.50

ESSENTIAL PORTUGUESE GRAMMAR, Alexander da R. Prista. vi + 114pp.
21650-0 Paperbound $1.35

ESSENTIAL SPANISH GRAMMAR, Seymour Resnick. 2500 word cognate list. 115pp.
(EBE) 20780-3 Paperbound $1.25

ESSENTIAL ENGLISH GRAMMAR, Philip Gucker. Combines best features of modern, functional and traditional approaches. For refresher, class use, home study. x + 177pp.
21649-7 Paperbound $1.75

A PHRASE AND SENTENCE DICTIONARY OF SPOKEN SPANISH. Prepared for U. S. War Department by U. S. linguists. As above, unit is idiom, phrase or sentence rather than word. English-Spanish and Spanish-English sections contain modern equivalents of over 18,000 sentences. Introduction and appendix as above. iv + 513pp.
20495-2 Paperbound $3.50

A PHRASE AND SENTENCE DICTIONARY OF SPOKEN RUSSIAN. Dictionary prepared for U. S. War Department by U. S. linguists. Basic unit is not the word, but the idiom, phrase or sentence. English-Russian and Russian-English sections contain modern equivalents for over 30,000 phrases. Grammatical introduction covers phonetics, writing, syntax. Appendix of word lists for food, numbers, geographical names, etc. vi + 573 pp. 6⅛ x 9¼.
20496-0 Paperbound $5.50

CONVERSATIONAL CHINESE FOR BEGINNERS, Morris Swadesh. Phonetic system, beginner's course in Pai Hua Mandarin Chinese covering most important, most useful speech patterns. Emphasis on modern colloquial usage. Formerly *Chinese in Your Pocket.* xvi + 158pp.
21123-1 Paperbound $1.75

How to Know the Wild Flowers, Mrs. William Starr Dana. This is the classical book of American wildflowers (of the Eastern and Central United States), used by hundreds of thousands. Covers over 500 species, arranged in extremely easy to use color and season groups. Full descriptions, much plant lore. This Dover edition is the fullest ever compiled, with tables of nomenclature changes. 174 full-page plates by M. Satterlee. xii + 418pp. 20332-8 Paperbound $3.00

Our Plant Friends and Foes, William Atherton DuPuy. History, economic importance, essential botanical information and peculiarities of 25 common forms of plant life are provided in this book in an entertaining and charming style. Covers food plants (potatoes, apples, beans, wheat, almonds, bananas, etc.), flowers (lily, tulip, etc.), trees (pine, oak, elm, etc.), weeds, poisonous mushrooms and vines, gourds, citrus fruits, cotton, the cactus family, and much more. 108 illustrations. xiv + 290pp. 22272-1 Paperbound $2.50

How to Know the Ferns, Frances T. Parsons. Classic survey of Eastern and Central ferns, arranged according to clear, simple identification key. Excellent introduction to greatly neglected nature area. 57 illustrations and 42 plates. xvi + 215pp. 20740-4 Paperbound $2.00

Manual of the Trees of North America, Charles S. Sargent. America's foremost dendrologist provides the definitive coverage of North American trees and tree-like shrubs. 717 species fully described and illustrated: exact distribution, down to township; full botanical description; economic importance; description of sub-species and races; habitat, growth data; similar material. Necessary to every serious student of tree-life. Nomenclature revised to present. Over 100 locating keys. 783 illustrations. lii + 934pp. 20277-1, 20278-X Two volumes, Paperbound $7.00

Our Northern Shrubs, Harriet L. Keeler. Fine non-technical reference work identifying more than 225 important shrubs of Eastern and Central United States and Canada. Full text covering botanical description, habitat, plant lore, is paralleled with 205 full-page photographs of flowering or fruiting plants. Nomenclature revised by Edward G. Voss. One of few works concerned with shrubs. 205 plates, 35 drawings. xxviii + 521pp. 21989-5 Paperbound $3.75

The Mushroom Handbook, Louis C. C. Krieger. Still the best popular handbook: full descriptions of 259 species, cross references to another 200. Extremely thorough text enables you to identify, know all about any mushroom you are likely to meet in eastern and central U. S. A.: habitat, luminescence, poisonous qualities, use, folklore, etc. 32 color plates show over 50 mushrooms, also 126 other illustrations. Finding keys. vii + 560pp. 21861-9 Paperbound $4.50

Handbook of Birds of Eastern North America, Frank M. Chapman. Still much the best single-volume guide to the birds of Eastern and Central United States. Very full coverage of 675 species, with descriptions, life habits, distribution, similar data. All descriptions keyed to two-page color chart. With this single volume the average birdwatcher needs no other books. 1931 revised edition. 195 illustrations. xxxvi + 581pp. 21489-3 Paperbound $5.00

AMERICAN FOOD AND GAME FISHES, David S. Jordan and Barton W. Evermann. Definitive source of information, detailed and accurate enough to enable the sportsman and nature lover to identify conclusively some 1,000 species and sub-species of North American fish, sought for food or sport. Coverage of range, physiology, habits, life history, food value. Best methods of capture, interest to the angler, advice on bait, fly-fishing, etc. 338 drawings and photographs. 1 + 574pp. 6⅝ x 9⅜.

22196-2 Paperbound $5.00

THE FROG BOOK, Mary C. Dickerson. Complete with extensive finding keys, over 300 photographs, and an introduction to the general biology of frogs and toads, this is the classic non-technical study of Northeastern and Central species. 58 species; 290 photographs and 16 color plates. xvii + 253pp.

21973-9 Paperbound $4.00

THE MOTH BOOK: A GUIDE TO THE MOTHS OF NORTH AMERICA, William J. Holland. Classical study, eagerly sought after and used for the past 60 years. Clear identification manual to more than 2,000 different moths, largest manual in existence. General information about moths, capturing, mounting, classifying, etc., followed by species by species descriptions. 263 illustrations plus 48 color plates show almost every species, full size. 1968 edition, preface, nomenclature changes by A. E. Brower. xxiv + 479pp. of text. 6½ x 9¼.

21948-8 Paperbound $6.00

THE SEA-BEACH AT EBB-TIDE, Augusta Foote Arnold. Interested amateur can identify hundreds of marine plants and animals on coasts of North America; marine algae; seaweeds; squids; hermit crabs; horse shoe crabs; shrimps; corals; sea anemones; etc. Species descriptions cover: structure; food; reproductive cycle; size; shape; color; habitat; etc. Over 600 drawings. 85 plates. xii + 490pp.

21949-6 Paperbound $4.00

COMMON BIRD SONGS, Donald J. Borror. 33⅓ 12-inch record presents songs of 60 important birds of the eastern United States. A thorough, serious record which provides several examples for each bird, showing different types of song, individual variations, etc. Inestimable identification aid for birdwatcher. 32-page booklet gives text about birds and songs, with illustration for each bird.

21829-5 Record, book, album. Monaural, $3.50

FADS AND FALLACIES IN THE NAME OF SCIENCE, Martin Gardner. Fair, witty appraisal of cranks and quacks of science: Atlantis, Lemuria, hollow earth, flat earth, Velikovsky, orgone energy, Dianetics, flying saucers, Bridey Murphy, food fads, medical fads, perpetual motion, etc. Formerly "In the Name of Science." x + 363pp.

20394-8 Paperbound $3.00

HOAXES, Curtis D. MacDougall. Exhaustive, unbelievably rich account of great hoaxes: Locke's moon hoax, Shakespearean forgeries, sea serpents, Loch Ness monster, Cardiff giant, John Wilkes Booth's mummy, Disumbrationist school of art, dozens more; also journalism, psychology of hoaxing. 54 illustrations. xi + 338pp.

20465-0 Paperbound $3.50

THE PRINCIPLES OF PSYCHOLOGY, William James. The famous long course, complete and unabridged. Stream of thought, time perception, memory, experimental methods—these are only some of the concerns of a work that was years ahead of its time and still valid, interesting, useful. 94 figures. Total of xviii + 1391pp.
20381-6, 20382-4 Two volumes, Paperbound $9.00

THE STRANGE STORY OF THE QUANTUM, Banesh Hoffmann. Non-mathematical but thorough explanation of work of Planck, Einstein, Bohr, Pauli, de Broglie, Schrödinger, Heisenberg, Dirac, Feynman, etc. No technical background needed. "Of books attempting such an account, this is the best," Henry Margenau, Yale. 40-page "Postscript 1959." xii + 285pp. 20518-5 Paperbound $3.00

THE RISE OF THE NEW PHYSICS, A. d'Abro. Most thorough explanation in print of central core of mathematical physics, both classical and modern; from Newton to Dirac and Heisenberg. Both history and exposition; philosophy of science, causality, explanations of higher mathematics, analytical mechanics, electromagnetism, thermodynamics, phase rule, special and general relativity, matrices. No higher mathematics needed to follow exposition, though treatment is elementary to intermediate in level. Recommended to serious student who wishes verbal understanding. 97 illustrations. xvii + 982pp. 20003-5, 20004-3 Two volumes, Paperbound $10.00

GREAT IDEAS OF OPERATIONS RESEARCH, Jagjit Singh. Easily followed non-technical explanation of mathematical tools, aims, results: statistics, linear programming, game theory, queueing theory, Monte Carlo simulation, etc. Uses only elementary mathematics. Many case studies, several analyzed in detail. Clarity, breadth make this excellent for specialist in another field who wishes background. 41 figures. x + 228pp. 21886-4 Paperbound $2.50

GREAT IDEAS OF MODERN MATHEMATICS: THEIR NATURE AND USE, Jagjit Singh. Internationally famous expositor, winner of Unesco's Kalinga Award for science popularization explains verbally such topics as differential equations, matrices, groups, sets, transformations, mathematical logic and other important modern mathematics, as well as use in physics, astrophysics, and similar fields. Superb exposition for layman, scientist in other areas. viii + 312pp.
20587-8 Paperbound $2.75

GREAT IDEAS IN INFORMATION THEORY, LANGUAGE AND CYBERNETICS, Jagjit Singh. The analog and digital computers, how they work, how they are like and unlike the human brain, the men who developed them, their future applications, computer terminology. An essential book for today, even for readers with little math. Some mathematical demonstrations included for more advanced readers. 118 figures. Tables. ix + 338pp. 21694-2 Paperbound $2.50

CHANCE, LUCK AND STATISTICS, Horace C. Levinson. Non-mathematical presentation of fundamentals of probability theory and science of statistics and their applications. Games of chance, betting odds, misuse of statistics, normal and skew distributions, birth rates, stock speculation, insurance. Enlarged edition. Formerly "The Science of Chance." xiii + 357pp. 21007-3 Paperbound $2.50

PLANETS, STARS AND GALAXIES: DESCRIPTIVE ASTRONOMY FOR BEGINNERS, A. E. Fanning. Comprehensive introductory survey of astronomy: the sun, solar system, stars, galaxies, universe, cosmology; up-to-date, including quasars, radio stars, etc. Preface by Prof. Donald Menzel. 24pp. of photographs. 189pp. 5¼ x 8¼.
21680-2 Paperbound $2.50

TEACH YOURSELF CALCULUS, P. Abbott. With a good background in algebra and trig, you can teach yourself calculus with this book. Simple, straightforward introduction to functions of all kinds, integration, differentiation, series, etc. "Students who are beginning to study calculus method will derive great help from this book." Faraday House Journal. 308pp.
20683-1 Clothbound $2.50

TEACH YOURSELF TRIGONOMETRY, P. Abbott. Geometrical foundations, indices and logarithms, ratios, angles, circular measure, etc. are presented in this sound, easy-to-use text. Excellent for the beginner or as a brush up, this text carries the student through the solution of triangles. 204pp.
20682-3 Clothbound $2.00

BASIC MACHINES AND HOW THEY WORK, U. S. Bureau of Naval Personnel. Originally used in U.S. Naval training schools, this book clearly explains the operation of a progression of machines, from the simplest—lever, wheel and axle, inclined plane, wedge, screw—to the most complex—typewriter, internal combustion engine, computer mechanism. Utilizing an approach that requires only an elementary understanding of mathematics, these explanations build logically upon each other and are assisted by over 200 drawings and diagrams. Perfect as a technical school manual or as a self-teaching aid to the layman. 204 figures. Preface. Index. vii + 161pp. 6½ x 9¼.
21709-4 Paperbound $2.50

THE FRIENDLY STARS, Martha Evans Martin. Classic has taught naked-eye observation of stars, planets to hundreds of thousands, still not surpassed for charm, lucidity, adequacy. Completely updated by Professor Donald H. Menzel, Harvard Observatory. 25 illustrations. 16 x 30 chart. x + 147pp.
21099-5 Paperbound $2.00

MUSIC OF THE SPHERES: THE MATERIAL UNIVERSE FROM ATOM TO QUASAR, SIMPLY EXPLAINED, Guy Murchie. Extremely broad, brilliantly written popular account begins with the solar system and reaches to dividing line between matter and nonmatter; latest understandings presented with exceptional clarity. Volume One: Planets, stars, galaxies, cosmology, geology, celestial mechanics, latest astronomical discoveries; Volume Two: Matter, atoms, waves, radiation, relativity, chemical action, heat, nuclear energy, quantum theory, music, light, color, probability, antimatter, antigravity, and similar topics. 319 figures. 1967 (second) edition. Total of xx + 644pp.
21809-0, 21810-4 Two volumes, Paperbound $5.75

OLD-TIME SCHOOLS AND SCHOOL BOOKS, Clifton Johnson. Illustrations and rhymes from early primers, abundant quotations from early textbooks, many anecdotes of school life enliven this study of elementary schools from Puritans to middle 19th century. Introduction by Carl Withers. 234 illustrations. xxxiii + 381pp.
21031-6 Paperbound $4.00

MATHEMATICAL PUZZLES FOR BEGINNERS AND ENTHUSIASTS, Geoffrey Mott-Smith. 189 puzzles from easy to difficult—involving arithmetic, logic, algebra, properties of digits, probability, etc.—for enjoyment and mental stimulus. Explanation of mathematical principles behind the puzzles. 135 illustrations. viii + 248pp.
20198-8 Paperbound $2.00

PAPER FOLDING FOR BEGINNERS, William D. Murray and Francis J. Rigney. Easiest book on the market, clearest instructions on making interesting, beautiful origami. Sail boats, cups, roosters, frogs that move legs, bonbon boxes, standing birds, etc. 40 projects; more than 275 diagrams and photographs. 94pp.
20713-7 Paperbound $1.00

TRICKS AND GAMES ON THE POOL TABLE, Fred Herrmann. 79 tricks and games— some solitaires, some for two or more players, some competitive games—to entertain you between formal games. Mystifying shots and throws, unusual caroms, tricks involving such props as cork, coins, a hat, etc. Formerly *Fun on the Pool Table.* 77 figures. 95pp.
21814-7 Paperbound $1.25

HAND SHADOWS TO BE THROWN UPON THE WALL: A SERIES OF NOVEL AND AMUSING FIGURES FORMED BY THE HAND, Henry Bursill. Delightful picturebook from great-grandfather's day shows how to make 18 different hand shadows: a bird that flies, duck that quacks, dog that wags his tail, camel, goose, deer, boy, turtle, etc. Only book of its sort. vi + 33pp. 6½ x 9¼. 21779-5 Paperbound $1.00

WHITTLING AND WOODCARVING, E. J. Tangerman. 18th printing of best book on market. "If you can cut a potato you can carve" toys and puzzles, chains, chessmen, caricatures, masks, frames, woodcut blocks, surface patterns, much more. Information on tools, woods, techniques. Also goes into serious wood sculpture from Middle Ages to present, East and West. 464 photos, figures. x + 293pp.
20965-2 Paperbound $2.50

HISTORY OF PHILOSOPHY, Julián Marías. Possibly the clearest, most easily followed, best planned, most useful one-volume history of philosophy on the market; neither skimpy nor overfull. Full details on system of every major philosopher and dozens of less important thinkers from pre-Socratics up to Existentialism and later. Strong on many European figures usually omitted. Has gone through dozens of editions in Europe. 1966 edition, translated by Stanley Appelbaum and Clarence Strowbridge. xviii + 505pp. 21739-6 Paperbound $3.50

YOGA: A SCIENTIFIC EVALUATION, Kovoor T. Behanan. Scientific but non-technical study of physiological results of yoga exercises; done under auspices of Yale U. Relations to Indian thought, to psychoanalysis, etc. 16 photos. xxiii + 270pp.
20505-3 Paperbound $2.50

Prices subject to change without notice.
Available at your book dealer or write for free catalogue to Dept. GI, Dover Publications, Inc., 180 Varick St., N. Y., N. Y. 10014. Dover publishes more than 150 books each year on science, elementary and advanced mathematics, biology, music, art, literary history, social sciences and other areas.